Labour's Immigratic

Erica Consterdine

Labour's Immigration Policy

The Making of the Migration State

palgrave
macmillan

Erica Consterdine
Department of Politics and Sussex Centre
for Migration Research
University of Sussex
Brighton, United Kingdom

ISBN 978-3-319-64691-6   ISBN 978-3-319-64692-3 (eBook)
https://doi.org/10.1007/978-3-319-64692-3

Library of Congress Control Number: 2017954209

© The Editor(s) (if applicable) and The Author(s) 2018
This work is subject to copyright. All rights are solely and exclusively licensed by the Publisher, whether the whole or part of the material is concerned, specifically the rights of translation, reprinting, reuse of illustrations, recitation, broadcasting, reproduction on microfilms or in any other physical way, and transmission or information storage and retrieval, electronic adaptation, computer software, or by similar or dissimilar methodology now known or hereafter developed.
The use of general descriptive names, registered names, trademarks, service marks, etc. in this publication does not imply, even in the absence of a specific statement, that such names are exempt from the relevant protective laws and regulations and therefore free for general use.
The publisher, the authors and the editors are safe to assume that the advice and information in this book are believed to be true and accurate at the date of publication. Neither the publisher nor the authors or the editors give a warranty, express or implied, with respect to the material contained herein or for any errors or omissions that may have been made. The publisher remains neutral with regard to jurisdictional claims in published maps and institutional affiliations.

Cover image: © fandijki / Getty

Printed on acid-free paper

This Palgrave Macmillan imprint is published by Springer Nature
The registered company is Springer International Publishing AG
The registered company address is: Gewerbestrasse 11, 6330 Cham, Switzerland

Preface

A decade ago, one of my former lecturers told me that if you really want to understand any given society, you should take a look at their immigration system. In fact, one of the protagonists to this unprecedented period of policymaking, Tony Blair, said something similar and that is indeed where we start the book. It was this realisation—that this specific policy arena can represent or mirror a society—that sowed the seeds of inspiration that led to this book. I believe such sentiments are truer today than when I first heard them 10 years ago.

Hardly a day goes by without immigration featuring in the headlines. The issue dominates debate across the political spectrum and has been a top voting issue amongst the British public for over a decade. It is one of the most divisive, contested and important issues of our time. The referendum in Britain on membership of the EU in June 2016 sent shockwaves across the political establishment and was undoubtedly shaped, possibly even determined, by the politics of immigration. Immigration has shifted from the periphery to the centre of the political landscape and will likely be a fixture in Britain for years to come. To understand how and why immigration has gravitated from low to high politics, we have to turn to the New Labour government's period in office between 1997 and 2010.

Under Labour, Britain's economic immigration policy went from a highly restrictive approach to one of the most liberal in Europe. Historically for Britain, and comparatively across Europe, Labour's reforms were an unprecedented policy reversal. With two and a half million foreign born workers added to the population since 1997, and over half of Britain's foreign born population arriving between 2001 and 2011, immigration under Labour quite literally changed the face of Britain. This period was, and is, the *Making of a Migration State*.

The *Making of a Migration State* explains why such a policy transformation transpired under the Labour governments by unpacking the mechanisms and processes that led to such an unexpected outcome. Ultimately, this book is about why governments liberalise economic immigration policy and the unintended consequences of intended actions. This book will be of interest for anybody who wants to understand why immigration is dominating the political debate and will be essential reading for those wanting to know why governments pursue expansive immigration regimes.

No (wo)man is an island, and there are too many people to mention to fully express my gratitude. I would like to thank the ESRC for financially supporting this research (grant number MP/21013669) as well as my colleagues at the Sussex European Institute and the Sussex Centre for Migration Research for their support, wisdom and encouragement. I am indebted to all my interviewees; they were all generous with their time and without their candidness this research would not have been possible. Four academics to whom I am enormously grateful—Adam Fishwick, Liam Stanley, Christina Boswell and Paul Taggart—have all provided much needed critical feedback at every stage. I would also like to thank the two anonymous peer reviewers, whose suggestions I believe have improved the book along with the editorial team at Palgrave who made the publication process swift and easy. My intellectual debt goes to Paul Statham and in particular James Hampshire who has been my mentor for almost a decade. I'm greatly indebted for all of his time, encouragement and his creative inspiration. Much needed support, laughter and love have come from my friends and family, in particular my parents Malcolm and Tracey. Finally, my eternal gratitude to my husband, Adam, who not only read every page of the book but gave me all the support I've ever needed over the last nine years. He knows far more about immigration policy than he ever wanted to.

Contents

1 Introduction: The Puzzle of Managed Migration   1

2 A Framework for Understanding Immigration Policy   21

3 Making the Migration State: The History of Britain's Immigration Policy   53

4 In Whose Interest? Organised Interests, Policy Networks and Collective Action   87

5 Do Parties Matter? Party Ideology and Party Competition   119

6 Bringing the State Back In: Institutional Change and the Administrative Context   161

7 An Unintended Consequence...   201

8   Beyond New Labour   219

Appendix 1: Overview of Economic Immigration Policy Reforms and Rhetoric, 1997–2010   243

Index   245

List of Abbreviations

A2	Two countries that joined EU in 2007: Bulgaria and Romania
A8	Eight countries that joined EU in 2004: Czech Republic, Estonia, Hungary, Latvia, Lithuania, Poland, Slovakia, Slovenia
BCC	British Chamber of Commerce
BHA	British Hospitality Association
BIS	Department for Business Innovation and Skills
CBI	Confederation for British Industries
CEE(C)	Central and Eastern European (Countries)
CIPD	Chartered Institute of Personnel and Development
CRE	Commission for Racial Equality
CV	Civil Servant
DCLG	Department for Communities and Local Government
Defra	Department for Environment, Food and Rural Affairs
DfEE	Department for Education and Employment
DTI	Department for Trade and Industry
DWP	Department for Work and Pensions
EBP	Evidence-based policy
EEC	European Economic Community
EU	European Union
FCO	Foreign and Commonwealth Office
GATS	General Agreement on Trades in Services
GSK	Glaxo Smithkline (pharmaceutical company)
HMRC	Her Majesty's Revenue and Customs
HSMP	Highly skilled migrant programme

ILPA	Immigration Law Practitioners Association
IND	Immigration and Nationality Directorate
IOD	Institute of Directors
IPPR	Institute for Public Policy Research
IRSS	Immigration Research and Statistics Service
JCWI	Joint Council for the Welfare of Immigrants
JUG	Joined-up Government
LPRAG	Labour Party Race Action Group
MAC	Migration Advisory Committee
MBA	Migration and Border Analysis
MIF	Migration Impacts Forum
MW	Migration Watch
NEC	National Executive Committee
NFU	National Farmers Union
NHS	National Health Service
PBS	Points-based system
PIU	Performance and Innovation Unit
PMI1(and 2)	Prime Minister's Initiative for International Education (1= 1999–2005; 2=2006–2011)
PSAs	Public Service Agreements
RLMT	Resident Labour Market Test
SAWS	Seasonal Agricultural Workers Scheme
SBS	Sector-based scheme
SpAd	Special Advisor
TGWU	Transport and General Workers' Union
TUC	Trade Union Congress
UK	United Kingdom
UKBA	UK Border Agency
UKCISA	UK Council for International Student Affairs
UUK	Universities UK
WRS	Workers Registration Scheme

List of Tables

Table 4.1　Collective action strategies　89
Table 4.2　Economic immigration policy network under New Labour　97

CHAPTER 1

Introduction: The Puzzle of Managed Migration

> *"I think it is a good rule of thumb to ask of a country: are people trying to get into it or out of it? It's not a bad guide to what sort of country it is"*
>
> (Tony Blair 2003)

1.1 Introduction

Hardly a day goes by without immigration featuring in the headlines. The issue dominates debate across the political spectrum and has been a top voting issue amongst the British public for over a decade (Duffy and Frere-Smith 2014; Blinder and Allen 2016), becoming the most important issue facing Britain for voters in 2014 (Dennison and Goodwin 2015, 173). It is one of the most divisive and at the same time, with public concern over immigration being acute amongst working and middle classes and across partisan divides (Fabian Society 2017), paradoxically unifying issue of our times.

The referendum in Britain on membership in the EU in June 2016 sent shockwaves across the political establishment not just in Britain itself but also throughout Europe and the world beyond. This was a campaign and, some would say, a vote fuelled by anti-migrant sentiment (Portes 2016). Current Prime Minister Theresa May is so convinced that 'Brexit must mean control of the number of people who come to Britain from Europe' (May 2017) that the government, against damaging economic forecasts, plan to take Britain out of the single market for the apparent trade-off of

© The Author(s) 2018
E. Consterdine, *Labour's Immigration Policy*,
https://doi.org/10.1007/978-3-319-64692-3_1

1

reduced immigration. Immigration has undoubtedly shifted from the periphery to the centre of the political landscape and will be a fixture in Britain for years to come. To understand how and why immigration has gravitated from low to high politics, we have to turn to the New Labour government's period in office between 1997 and 2010.

Under New Labour, Britain's economic (or labour) immigration policy went from a highly restrictive approach to one of the most expansive in Europe: work permit criteria were relaxed, international students were doubled, the government expanded existing and launched new low and high skilled migrant worker schemes, and, from 2005, a new points-based system (PBS) was initiated. Overshadowing these important reforms was the decision in 2004 to allow citizens of the eight EU accession states the right to work in Britain, resulting in one of the largest migration flows in Britain's peacetime history. Couched in the narrative of managed migration, these policy reforms signified a new approach to immigration based on economic utilitarian arguments (Balch 2010). Coupled with the mantra of attracting the 'brightest and best' immigrants, managed migration denoted an alternative immigration system based on the supply and demand of skills, and above all embracing the positive economic benefits of immigration. With two and a half million foreign born workers added to the population since 1997, and over half of Britain's foreign born population arriving between 2001 and 2011 (ONS 2012), immigration under Labour 'quite literally changed the face of Britain' (Finch and Goodhart 2010, 3). This period was, and is, the *Making of a Migration State*.

New Labour's managed migration policy stood in stark contrast to Britain's restrictive immigration past. Writing in 1994, Gary P. Freeman famously described Britain as a 'deviant case' in Western European migration policy. For over three decades, successive British governments had managed to combine a liberal approach to flows of capital and trade with effective limits on the flow of immigrants. Historically for Britain, and comparatively across Europe, Labour's reforms were an 'unprecedented policy reversal' (Hansen 2014).

By the time Labour left office in 2010 then, a 'reluctant country of immigration' (Layton-Henry 1994) had been transformed into a fully-fledged 'migration state' (Hollifield 2004). This was the defining breakpoint between Britain's post-war bipartisan consensus of 'zero immigration' (Freeman 1994) and today's political fixture, where far from being a taboo subject for politicians, immigration could not figure more prominently in political debate. This was an unprecedented period of immigration policymaking, which both broke with the past and set the stage for where Britain is now.

The Labour government's rapid policy change is puzzling for at least two reasons. First, the existing political science literature has often emphasised the 'path dependent' character of immigration policy in Britain and indeed elsewhere (Hansen 2000; Tichenor 2002), suggesting that immigration policy change is likely to be incremental at most. Immigration policy is often shaped by legacies of the past because policies can change populations and set the policy norms for successive administrations (Ellerman 2015; Wright 2012). Second, in no Western country can a party gain votes by promoting or expanding immigration (Lahav 1997). The Labour government's liberalisation of immigration policy went against public opinion, and therefore there was no obvious electoral dividend to their expansive regime. Whilst the British public has long been in favour of reducing immigration, the high level of public concern began in 2000, at a time where the New Labour government were pursuing the most expansive immigration regime to date (Ipsos Mori 2015; Evans and Chzhen 2013). Indeed, Labour's policies were certainly not a vote winner; they have since dogged Labour's time in opposition, and public concern about large-scale immigration contributed to their electoral defeat in 2010 (Carey and Geddes 2010; Bale 2014) and hindered their chances of winning office in 2015 (Beckett 2016, 7; Cruddas 2016; Geddes and Tonge 2015).

How then to explain a change that was both electorally risky and ran counter to Britain's past immigration policy? How did a country that was defined by its 'aspiration for zero immigration' (Freeman 1994) evolve into a fully fledged 'migration state'? The *Making of a Migration State* explains why such a policy transformation transpired under the Labour governments by unpacking the mechanisms and processes that led to such an unexpected outcome. Ultimately, this book is about why governments liberalise economic immigration policy and the unintended consequences of intended actions. This book will be of interest for anybody who wants to understand why immigration is dominating the political debate and will be essential reading for those wanting to know why governments pursue expansive immigration regimes.

1.2 Unpacking the Migration State

The objective of this book is to explain the expansionary developments of economic immigration policy under the Labour administrations of 1997–2010. It is important to stress from the outset that the focus of this research is explicitly with labour and student immigration, which combined I refer to hereon in as economic immigration. These two categories are closely related because these streams are 'wanted' immigration in contrast to 'unwanted' immigration such as irregular, humanitarian or family migration (Joppke 1998). I use the terms expansive and/or liberal policy to signify the Labour government's approach to facilitate entry of migrant workers, rather than any liberalisation in terms of migrant rights. Although workers rights in terms of transitions and qualifying settlement periods were also loosened under Labour in conjunction with the wider managed migration regime. While other areas of immigration policy, such as asylum and irregular immigration, became increasingly restrictive during this period, the Labour government's economic immigration policy, which this book is concerned with, was undoubtedly an expansive one.

In the context of economic globalisation and an embedded international human rights discourse, some contend that there has been a decline in power, significance and sovereignty of the nation state. In turn, it is argued that nation states have 'lost control' of their borders and are thus no longer the crucial actors in immigration policymaking (Soysal 1994; Sassen 1996; Jacobson 1996). This may hold true for some migration streams, such as humanitarian immigration or family reunification where international conventions can override domestic autonomy, but given that the nation state primarily determines the management and regulation of economic immigration policy, at least in Britain, this book employs approaches that focus on the domestic political arena.

The literature on immigration was once dominated by accounts from economists and sociologists suggesting (if only tacitly) that the nation-state and the institutions that comprise it were of secondary importance relative to international market forces and the personal networks that drive individuals to migrate (Castles 2004; Wright 2010). Yet what 'governments do matters a great deal' (Castles and Miller 2003, 94) in terms of explaining migratory movements. While immigration flows are not entirely determined by states, the decision 'to accept or reject aliens has not been relegated to actors other than the state, and the infrastructural capacity of modern states has not decreased, but increased over time' (Joppke 1998, 267). Independently of other conditions, 'it is state actions with respect to borders that determine whether any international migration will take place' (Zolberg 1989, 205). In other words, without nation-states and their associated apparatuses that define their borders, 'there would be no such thing as international migration' (Balch 2010, 4). The state still retains an active role in defining how liberal or restrictive an immigration regime is, and it is the state that ultimately decides who enters and resides legally, naturalises and can become part of the nation (Guiraudon and Lahav 2000, 167).

This is not to say that policy is the only driver of immigration or that policies always achieve their intended outcomes (Cornelieus et al. 1994). For example, the sharp increase in net migration in Labour's first year of office (1997–1998) was largely beyond the government's control and attributable to other factors. Net migration rose from 48,000 in 1997 to 140,000 in 1998 in large part because of: a rise in asylum applications following the Kosovo War (Home Office 1999), emigration decreasing from 45,000 to 11,000 (ONS 2006, 13) and EU immigration from the 15 Member States rising from 18,000 in 1997 to 33,000 in 1998 (ONS 2014). Thus net migration rose because of both an increase in inflow of 63,000 and a 28,000 reduction in emigration, neither of which in this case was due to any policy action by the Labour government. However, whilst the economic and social push and pull factors that drive immigration explain some of the increase in net migration under Labour, it is fair to say that the unprecedented increase overall was largely due to the government's policy reforms, in particular the A8 decision in 2004.

Policy levers do not always drive immigration flows then, and this is true across liberal states. Immigration policies in labour-importing countries have been said to be converging (Cornelieus et al. 2006), partly because of the shared challenges they face because of the liberal paradox. The liberal paradox first coined by James Hollifield (2004) refers to the contradictory pressures that the nation-state face on immigration, between market forces pushing states towards greater openness and powerful domestic pressures pushing towards closure. Alternatively, as James Hampshire puts it (2013, 12), there are the contradictory pulls between the logic of openness (because the liberal state is conditioned by constitutionalism and capitalism) and the logic of closure (because representative politics and nationhood are also facets of the liberal state). This leads Hollifield (2004) to conclude that 'trade and migration are inextricably linked…Hence, the rise of the trading state necessarily entails the rise of the migration state where considerations of power and interest are driven as much by migration as they are by commerce and finance' (p. 193). Whilst economic and social forces are the necessary pre-conditions for migration to occur, the 'sufficient conditions are legal and political' (Hollifield et al. 2008, 341) because 'states must be willing to accept immigration and to grant rights to outsiders' (Hollifield 2004, 885) in the first instance; thus policies themselves clearly have a significant role in shaping patterns and flows of immigration (Meyers 2000). It makes sense then that we would seek to understand the factors that drive states to formulate their immigration policies in the way they do.

If Britain was previously 'a country of zero immigration (Freeman 1994), it is hard to deny that the Labour administrations transformed Britain into a migration state in terms of numbers alone, even more so because, as we shall see, their reforms were driven by a capitalist imperative to some extent. The 2011 Census showed that the population of England and Wales was 56.1 million, a growth of 3.7 million or a 7.1 per cent increase in the 10 years since the last census in 2001, a period almost entirely governed by Labour. Fifty-six per cent of the population increase in England and Wales was due to migration. In the UK as a whole, the foreign-born population nearly doubled between 1993 and 2011 from 3.8 million to over 7 million, with almost 40 per cent of the foreign-born population arriving in 2004. This was the largest growth in the population in England and Wales in any 10-year period since the census began in 1801 (Duffy and Frere-Smith 2014).

This book is about explaining how and why a government expands labour immigration policy and the consequences of doing so in terms of the politics such a policy can produce. I therefore want to expand the concept of the Migration State to also refer to the politicisation of migration as a further component. What I mean by this is the way in which immigration has come to dominate the political debate, shaping voting intentions and becoming a contested policy arena across partisan divides. This is harder to quantify but nonetheless that immigration has gravitated from a marginal issue of concern to one of the top three voting issues in itself demonstrates how politicised immigration has become (Ipsos Mori 2015; Duffy 2014). Few would deny that immigration has become highly salient amongst the public, and, in turn, the policy dilemmas of immigration have consumed political elites.

The degree of saliency and polarization conditions whether an issue is politicised or not. Drawing from van der Brug et al. (2015, 6) it is the combination of agenda-orientated and conflict-orientated approaches that configure whether an issue is politicised. The agenda-setting literature (Jones and Baumgartner 2004; Kingdon 1995) tells us that it is only when a social topic is defined as a problem that we can really speak of a political issue. Agenda-setting theory focuses on the different thresholds that prevent a topic from becoming a political issue. It serves to reinforce that as long as 'the topic is treated as one that does not require state action, it is not politicised; it is not even a political issue' (van der Brug et al. 2015, p. 7). Although public concern over immigration was acute in early 1970s Britain (Saggar 2003), and so-called bogus asylum seekers received a vast amount of press attention in the late 1990s (Kaye 2001), these waves of public discontent are marginal in contrast to how much salience, or at least the importance, the public attribute to the issue now. As we shall see in Chap. 3, in the post-war period a bipartisan consensus of limited immigration dominated the political spectrum, which served to defuse the issue altogether so that immigration was not on the political agenda. In the 2010s increasingly elites frame immigration as a problem that requires state action. Immigration in 2010s Britain certainly fulfils the criteria of heightened saliency, intensified attention and resoundingly framed as a problem needing state action.

An issue only qualifies as politicised if there is also a high degree of conflict, be this conflict over the policy direction or conflict upon the means, and instruments to resolve the problem. The polarisation element of politicisation draws from the party politics or electoral competition school of thought (Downs 1957), scholars of which highlight the importance of positional competition and the extent to which different parties have polarising positions on the issue. When political actors have different positions on an issue they are in conflict, and thus the issue is polarised. Opposing positions may have always existed, but if the issue is not on the political agenda, the conflict is dormant (van der Brug et al. 2015, 5).

Conflict can divide both across and within parties. Thus where an issue produces intense intra-party conflict, parties and governments will try to de-emphasise issues on which they internally disagree. But of course parties and governments do not exercise full control over the political agenda. Newer parties, such as green parties politicising environmental issues or more aptly here radical right parties politicising immigration, often politicise new issues. Immigration in Britain clearly meets the criteria of a high-conflict and thus polarising issue, both across parties and perhaps more interestingly within parties. We delve into more detail on the politicisation of immigration in the epilogue chapter. Suffice to say for now that immigration in 2010s Britain is both highly salient and highly polarising and thus a politicised issue. These three components then—immigration being tied to trade, an actual unprecedented increase in net migration and immigration being highly salient, highly polarising and thus politicised—comprise Britain as a migration state.

Britain is of course not alone in becoming a migration state; as mentioned, liberal states are arguably converging on immigration policy. Wide ranges of explanations have been advanced to explain such convergence but these often overlook the policy process itself. Analysis of immigration in political science has been particularly attentive to the challenges immigration poses to the nation state (Joppke 1998; Hampshire 2013), but very few scholars have attempted to unpack the 'black box' of immigration policymaking. This has meant that the existing literature tends to focus on how just one aspect shapes policy outcomes. Whilst political economists look to the role of trade, production, economic context and demands from employers (Caviedes 2010; Freeman 2006; Cerna 2009; Menz 2008), institutionalists have demonstrated how liberal norms and international courts facilitate humanitarian and family migration (Guiraudon 2000; Joppke 1998). Meanwhile party politics scholars have shown us how mainstream and particularly extreme parties mobilise the issue (Bale et al. 2010; Mudde 2007: Norris 2005), in contrast to public opinion researchers who demonstrate the drivers of public anti-migration sentiments (Citrin et al. 1997; Ivarsflaten 2005; McLaren and Johnson 2007). Finally, scholars of national identity have shown how nation building, national cultures and policy legacies (Wright 2012; Ellerman 2015) can shape policy.

1.3 Multiple Lenses

Immigration policymaking has long been an explanatory challenge for political scientists because a myriad of factors and considerations shape policy outputs. As Jupp (1993, 254) notes, there is 'no single "scientific" analysis that is likely to provide a complete model for the politics of immigration policy', because any comprehensive analysis of immigration policymaking needs to consider a variety of determinants. Precisely because a number of factors shape immigration policy, to understand the policymaking process it makes sense to adopt a multiple-lens framework that can explain the different determinants of government action and elite preferences.

At any one time, governments must grapple with public demands and electoral competition, the needs of the labour market and the consequential demands from employers, conflicting policy visions from different departments, as well as geopolitical pressures and international conventions that restrict the autonomy of the state. Three approaches—organised interests, party politics and historical institutionalism—were used to understand the divergent objectives, drivers and considerations that influence the construction of immigration policy. By employing different lenses to the question of policy change, the book offers an account that recognises the multifaceted considerations of policymakers as well as the complexity of the policymaking process.

The first approach—organised interests—looks to the role of non-governmental actors to explain policy change. The organised interests approach posits that immigration policy is a product of bargaining and compromise between government and interest groups. This position contends that governments have expanded economic immigration because organised interests try to force governments to adopt specific policies (Freeman 1995; Menz 2008; Caviedes 2010). Central to this is the recognition that contemporary liberal states are capitalist states and are thus responsive to the demands of business (Hollifield 2004; Hampshire 2013).

Gary Freeman commented over 30 years ago that migrant labour was 'not merely a temporary convenience or necessity, but a structural requirement of advanced capitalism' (Freeman 1979, 3), and this remains the case. In lower-wage sectors migrants fill labour market shortages, in particular the so-called 3D jobs (dirty, dangerous or degrading), which indigenous populations are reluctant to do. At the other end of the scale, high-skilled immigrants have become imperative to fill skill shortages. The need for a mobile and flexible labour pool has intensified and, in a globalised economy, where large transnational firms operate without borders, greater flows of intra-corporate transfers are inevitable. The dependence on immigration has also extended to the higher education sector, where non-EU students and their considerable tuition fee contributions are now integral to financing the system as a whole. In essence, 'advanced capitalist states cannot afford—literally as well as metaphorically—not to solicit immigrants' (Hampshire 2013, 12).

The organised interest approach argues then that interest groups, especially employers in the case of economic immigration, will attempt to convince governments of the need for foreign labour and thus lobby governments for more expansive immigration policies. Accordingly, because these groups have more resources and are better organised than anti-migrant groups, governments respond to such mobilised demands as it is in their electoral interest to do so. Immigration policies are thus said to mirror the interests of those who can mobilise most effectively and/or have the most resources, and these tend to be those who stand to gain from expansive policies.

While those who adhere to the organised interests approach place primacy on the role of non-state actors, other authors contend that it is political parties that shape policy (Triadafilopoulos and Zaslove 2006; Bale 2008; Givens and Luedkte 2005). Stemming from a broadly elitist perspective, proponents of the '"politics matter" school of thought' (Imbeau et al. 2001, 1) argue that it is the political parties and the actors that comprise them that shape the political debate and ultimately determine policy. Political parties influence public policy by both translating public opinion into policies in exchange for support and at times acting as agents of change on the basis of ideologies (Schmidt 1996, 155). This perspective argues then that immigration policy is a product of partisan differences and/or party strategy.

Whilst parties are office-seeking organisations they are also fundamentally configured by a set of defining ideas (in other words a party ideology), which provides a coherent package of principles and beliefs. This ideology, in principle, reflects both the party's tenets and their core constituents concerns and thus acts as blueprint to guide party action. Immigration is an ideologically divisive issue for established parties of the left and right as it 'cuts across normal lines of political battle', precisely because it relates to wider social issues such as 'law and order, integration, employment and national identity' (Lahav 1997, 382; Freeman 1979; Triadafilopoulos and Zaslove 2006, 32). Nonetheless, party ideology and the broad left/right spectrum persist in shaping elite preferences to immigration to some degree (Ireland 2004; Lahav 1997), and fundamentally while office-seeking may be the primary motivation for parties to change policy, the essence of party success entails an 'achievement of a satisfactory trade-off between ideological introversion and electoral extroversion, between principles and power' (Bale 1999, 7).

Essentially parties matter in explaining policy change because unlike organised interests 'parties actually control the government' (Burstein and Linton 2002, 385). If party ideology conditions political elite preferences, since it is these actors that ultimately direct and enact policy, we would expect this to be reflected in policy. Accordingly, it is political parties and the elites that constitute them that condition the direction of immigration policy by way of channelling their ideology through policies, reflecting electoral preferences, competing with opposition parties to win voters and in turn structuring the political debate on immigration.

In contrast to those who focus on party political interactions, a third perspective looks to the state itself and the institutions that comprise it to explain government decisions on immigration policy. New institutionalism 'brings the state back in' by focusing on how administrations and bureaucracies shape immigration policy. The new institutionalist school claims that political institutions can be autonomous, and it is therefore these apolitical (in partisan political terms) institutions that form immigration policy according to the interests of the state. These emphasise the way in which actions of individuals exist within the context of the rules and norms of institutions (March and Olsen 2006). According to Boswell (2007, 79) there are two features of a definition of the state necessary for a neo-institutionalist analysis. The first is that the state cannot be understood as a monolithic entity; there must be some disaggregation between a system of party politics and the administration or the state's bureaucratic apparatus that determine the implementation of policy. Second, there must be conceptual space that allows for the possibility of the state having 'preferences which are not reducible to some matrix of societal interests' (Boswell 2007, 79). The autonomy of preferences could stem from the interest of the administration in securing legitimacy and/or the organisational dynamics and interests of different state departments.

One variant of new institutionalism, historical institutionalism, is particularly pertinent for explaining state decisions with regards to immigration policy, because immigration policy is so often a product of policy legacies (Wright 2012) and made and framed on the basis of long-held, embedded ideas about the objectives and ideal approach to the regulation of immigration (Hay 2006; Béland 2005; Hansen 2000). This approach argues that immigration policy is often a product of past political decisions—in terms of both the ideas structuring policy and more technical elements such as legislation—which constrains future action and thus creates a path-dependent effect on policy (Hansen 2000).

Drawing from a 'cultural' understanding of human agency (Hall and Taylor 1996; Hay and Wincott 1998), historical institutionalism postulates that the agencies, bureaucracies and departments which make immigration policy, are built on ideational foundations—that is, the initial construction of institutions are built on ideas. Subsequently these institutions develop ideas and framings on the policy areas within their remit in an autonomous manner, 'screened from political pressure' (Boswell 2007, 83). Through processes of normalisation and socialisation, certain ideas and framings become embedded in these institutions, which serve as cognitive filters through which actors come to interpret their environment. Crucially, the historical institutionalist is concerned with how, under certain conditions, such institutionalised ideas and paradigms, such as an established policy frame of immigration, are contested, challenged and replaced (Hay 2006, 65; Berman 1998). For the historical institutionalist, it is the state and the institutions that comprise it, which shape immigration policy.

In some ways these approaches offer 'self-contained "worlds" from which to view the policy process' (John 1998, 16), although they can work to complement each other. Organised interests focus on the associational relationships between non-state actors and government. The politics matter school of thought places primacy on the party composition of government and the preferences and interests of party actors. Historical institutionalists examine the norms and habits of policymaking in different policy sub-systems. Although these approaches are in 'dialogue with each other', they are also 'self-referential paradigms based on assumptions about the possibilities of human agency, the effect of structures, the meaning of power and the nature of the state' (John 1998, 17). Fundamentally each approach assumes that a different set of actors dominate and control the policy agenda—non-governmental actors, political parties and institutional actors respectively—and each stresses different causal mechanisms at play in policy change. The book applies each approach to the case of immigration policy under the Labour governments and examines the explanatory power they hold.

1.4 Plan of the Book

The book is organised as follows. Chapter 2 considers the three theoretical approaches adopted in more detail and addresses some key issues of defining policy change. The chapter explores the core explanatory argument of each approach and reviews the evidence in terms of how each factor has been demonstrated to shape immigration policy and gives details on the adopted methodology and research design. The chapter delves into how different sets of actors, including non-governmental actors, political parties and civil servants, are said to influence policy and establishes the conceptual and analytical tools to examine how interests, ideas and institutions can prompt policy change.

In Chapter 3 we move to Britain's history of immigration policy from the post-war period until Labour left office in 2010. Britain has long been a 'reluctant country of immigration' (Layton-Henry 1994) and given that the majority of Britain's post-war restrictive measures were targeted at non-white immigrants, many scholars contend that Britain's immigration regime was underpinned by racialisation. This chapter traces Britain's immigration evolution from a 'country of zero immigration' (Freeman 1994) to a migration state to illustrate the unprecedented shift under the Labour governments in comparison to Britain's post-war restrictive framing.

Chapter 4 turns to the role of non-state actors in Labour's immigration policy. Taking an organised interests lens the chapter examines whether policy change was a result of interest groups lobbying the government for expansive policies. The chapter explores the ways in which non-state actors, such as employers and employer associations, unions, sectoral interest groups and think tanks, did or did not influence immigration policy in the period 1997–2010.

The focus of Chapter 5 is on the governing party and the elites that comprised it. Here we consider the party political context and explore whether party ideology, intra-party change and party competition shaped immigration policy in this period. The chapter focuses on how the ideas of the governing party changed and the impact this had on the immigration policy preferences of the political elite. Relatively few scholars have examined how party ideology shapes immigration policy (see Odmalm 2014; Hinnfors et al. 2012 for exceptions), and this research seeks to fill this gap by analysing how Labour's ideology changed the preferences of the leading political elite.

Chapter 6 adopts a 'culture' understanding of historical institutionalism and considers the role of government departments and the policymaking process itself. The chapter does this by examining the administrative context of immigration policymaking, analysing the processes of policymaking, such as which departments were involved, how immigration was framed, and how entrenched institutional cultures did or did not influence immigration policy. The chapter also considers whether changes to the machinery of government—including an initiative for joined-up government and evidence-based policy—had an impact on immigration policymaking. This chapter delves into the 'black box' of policymaking and gives insights into the conflicting objectives that inform government decisions on economic immigration policy.

Chapter 7 brings the key arguments of the book together by summarising the findings from the three empirical chapters and reflecting on the utility of the different theoretical approaches employed for explaining policy change. The chapter calls for complex causality and provides an overarching explanation for this case of policy change, outlining the necessary conditions, ideas and context which gave way to a shift in the policy framing of immigration.

The final epilogue chapter reflects on New Labour's legacies on the politics of immigration and beyond. The chapter looks at the feedback effects of Labour's reforms on policymaking and the repercussions of Labour's policies on the broader political landscape to unravel how immigration has become so dominant in debate. New Labour's policies brought immigration to the fore of the political landscape, and in this sense, it is no exaggeration to say that the period under Labour has transformed the politics of immigration in Britain.

References

Balch, A. (2010). *Managing Labour Migration in Europe: Ideas, Knowledge and Policy Change*. Manchester: Manchester University Press.

Bale, T. (1999). *Sacred Cows and Common Sense: The Symbolic Statecraft and Political Culture of the British Labour Party*. Aldershot: Ashgate Publishing.

Bale, T. (2008). Turning Round the Telescope. Centre-right Parties and Immigration and Integration Policy in Europe. *Journal of European Public Policy, 15*(3), 315–330.

Bale, T. (2014). Putting It Right? The Labour Party's Big Shift on Immigration Since 2010. *The Political Quarterly, 85*(3), 296–303.

Bale, T., Green-Pedersen, C., Krouwel, A., Luther, K. R., & Sitter, N. (2010). If You Can't Beat Them, Join Them? Explaining Social Democratic Responses to the Challenge from the Populist Radical Right in Western Europe. *Political Studies, 58*(3), 410–426.

Beckett, M. (2016). *Labour Party: Learning the Lessons from Defeat Taskforce Report*. London: Labour Party.

Béland, D. (2005). Ideas and Social Policy: An Institutionalist Perspective. *Social Policy and Administrative, 39*(1), 1–18.

Berman, S. (1998). *The Social Democratic Moment: Ideas and Politics in the Making of Interwar Europe*. Cambridge: Harvard University Press.

Blair, T. (2003, January 7). *Prime Minister's Address to British Ambassadors*. London. Available from: http://www.theguardian.com/politics/2003/jan/07/foreignpolicy.speeches. Accessed on 10 Mar 2011.

Blinder, S., & Allen, W. (2016). *UK Public Opinion Towards Immigration: Overall Attitudes and Levels of Concern*. Oxford: Migration Observatory. http://www.migrationobservatory.ox.ac.uk/resources/briefings/uk-public-opinion-toward-immigration-overall-attitudes-and-level-of-concern/. Accessed 24 Feb 2017.

Boswell, C. (2007). Theorizing Migration Policy: Is There a Third Way? *International Migration Review, 41*(1), 75–100.

Burnstein, P., & Linton, A. (2002). The Impact of Political Parties, Interest Groups and Social Movement Organizations on Public Policy: Some Recent Evidence and Theoretical Concerns. *Social Forces, 81*(2), 341–408.

Carey, S., & Geddes, A. (2010). Less Is More: Immigration and European Integration at the 2010 General Election. *Parliamentary Affairs, 63*(4), 849–865.

Castles, S. (2004). The Factors That Make and Unmake Immigration Policies. *International Migration Review, 38*(3), 852–844.

Castles, S., & Miller, M. (2003). *The Age of Migration: International Population Movements in the Modern World*. Basingstoke: Palgrave Macmillan.

Caviedes, A. A. (2010). *Prying Open Fortress Europe: The Turn to Sectoral Labor Migration*. Lanham/Plymouth: Lexington Books.

Cerna, L. (2009). The Varieties of High-Skilled Immigration Policies: Coalitions and Policy Outputs in Advanced Industrial Countries. *Journal of European Public Policy, 16*(1), 144–161.

Citrin, J., Green, D. P., Muste, C., & Wong, C. (1997). Public Opinion Toward Immigration Reform: The Role of Economic Motivations. *The Journal of Politics, 59*(3), 858–881.

Cornelius, W., Tsuda, T., Martin, P., & Hollifield, J. (1994). *Controlling Immigration: A Global Perspective*. Stanford: Stanford University Press.

Cornelius, W., Tsuda, T., Martin, P., & Hollifield, J. (2006). *Controlling Immigration: A Global Perspective*. Stanford: Stanford University Press.

Cruddas, J. (2016). What We Can Learn from Labour's Crushing Election Defeat. *Labourlist* [Online]. http://labourlist.org/2016/05/labours-future-what-we-can-learn-from-the-election-loss/. Accessed on 13 Jan 2017.

Dennison, J., & Goodwin, M. (2015). Immigration, Issue Ownership and the Rise of UKIP. *Parliamentary Affairs, 68*(suppl 1), 168–187.

Downs, A. (1957). *An Economic Theory of Democracy*. New York: Harper.

Duffy, B. (2014). Perceptions and Reality: Ten Things We Should Know About Attitudes to Immigration in the UK. *The Political Quarterly, 85*(3), 259–266.

Duffy, B., & Frere-Smith, T. (2014). *Perceptions and Reality: Public Attitudes to Immgiration*. London: Ipsos Mori. https://www.ipsos-mori.com/DownloadPublication/1634_sri-perceptions-and-reality-immigration-report-2013.pdf. Accessed 13 Dec 2016.

Ellermann, A. (2015). Do Policy Legacies Matter? Past and Present Guest Worker Recruitment in Germany. *Journal of Ethnic and Migration Studies, 41*(8), 1235–1253.

Evans, G., & Chzhen, K. (2013). Explaining Voters' Defection from Labour Over the 2005–10 Electoral Cycle: Leadership, Economics and the Rising Importance of Immigration. *Political Studies, 61*(1 suppl), 138–157.

Fabian Society. (2017). *Stuck: How Labour Is Too Weak to Win, and Too Strong to Die*. London: Fabian Society [Online]. Available from http://www.fabians.org.uk/wp-content/uploads/2016/12/Stuck-Fabian-Society-analysis-paper.pdf. Accessed on 31 Jan 2017.

Finch, T., & Goodhart, D. (2010). Introduction. In T. Finch & D. Goodhart (Eds.), *Immigration Under Labour* (pp. 3–10). London: IPPR.

Freeman, G. (1979). *Immigrant Labor and Racial Conflict in Industrial Societies: The French and British Experience, 1945–1975*. Princeton: Princeton University Press.

Freeman, G. (1994). Commentary. In W. Cornelius, T. Tsuda, P. Martin, & J. Hollifield (Eds.), *Controlling Immigration: A Global Perspective* (pp. 297–303). Stanford: Stanford University Press.

Freeman, G. (1995). Modes of Immigration Politics in Liberal Democratic States. *International Migration Review, 29*(4), 881–902.

Freeman, G. (2006). National Models, Policy Types, and the Politics of Immigration in Liberal Democracies. *West European Politics, 29*(2), 227–247.

Geddes, A., & Tonge, J. (2015). *Britain Votes 2015*. Oxford: Oxford University Press.

Givens, T., & Luedtke, A. (2005). European Immigration Policies in Comparative Perspective: Issue Salience, Partisanship and Immigrant Rights. *Comparative European Politics, 3*(1), 1–22.

Guiraudon, V. (2000). European Integration and Migration Policy: Vertical Policy-Making as Venue Shopping. *JCMS: Journal of Common Market Studies, 38*(2), 251–271.

Guiraudon, V., & Lahav, G. (2000). A Reappraisal of the State Sovereignty Debate the Case of Migration Control. *Comparative Political Studies, 33*(2), 163–195.

Hall, P. A., & Taylor, R. C. (1996). Political Science and the Three New Institutionalisms. *Political Studies, 44*(5), 936–957.

Hampshire, J. (2013). *The Politics of Immigration*. Cambridge: Polity Press.

Hansen, R. (2000). *Citizenship and Immigration in Post-war Britain*. Oxford: Oxford University Press.

Hansen, R. (2014). Great Britain. In J. Hollifield, P. L. Martin, & P. M. Orrenius (Eds.), *Controlling Immigration: A Global Perspective* (pp. 199–220). Stanford: Stanford Press.

Hay, C. (2006). Constructivist Institutionalism. In R. A. W. Rhodes, S. A. Binder, & B. A. Rockman (Eds.), *The Oxford Handbook of Political Institutions* (pp. 56–75). Oxford: Oxford University Press.

Hay, C., & Wincott, D. (1998). Structure, Agency and Historical Institutionalism. *Political Studies, 46*(5), 951–957.

Hinnfors, J., Spehar, A., & Bucken-Knapp, G. (2012). The Missing Factor: Why Social Democracy Can Lead to Restrictive Immigration Policy. *Journal of European Public Policy, 19*(4), 1–19.

Hollifield, J. (2004). The Emerging Migration State. *International Migration Review, 38*(3), 885–912.

Hollifield, J. F., Hunt, V. F., & Tichenor, D. J. (2008). The Liberal Paradox: Immigrants, Markets and Rights in the United States. *SMU Law Review, 61*, 67–98.

Home Office. (1999). ASYLUM STATISTICS UNITED KINGDOM 1998 *By Madeleine Watson and Rod McGregor. Web Archives.* http://webarchive.nationalarchives.gov.uk/20110218135832/http://rds.homeoffice.gov.uk/rds/pdfs/hosb1099.pdf. Accessed 14 Feb 2017.

Imbeau, L. M., Pétry, F., & Lamari, M. (2001). Left–Right Party Ideology and Government Policies: A Meta–Analysis. *European Journal of Political Research, 40*(1), 1–29.

Ipsos-Mori. (2015). *Issues Facing Britain: August 2015 Issues Index.* Available from: https://www.ipsos-mori.com/researchpublications/researcharchive/3614/EconomistIpsos-MORI-August-2015-Issues-Index.aspx. Accessed 15 Sept 2015.

Ireland, P. (2004). *Becoming Europe: Immigration, Integration and the Welfare State.* Pittsburgh: University of Pittsburgh.

Ivarsflaten, E. (2005). Threatened by Diversity: Why Restrictive Asylum and Immigration Policies Appeal to Western Europeans. *Journal of Elections, Public Opinion & Parties, 15*(1), 21–45.

Jacobson, D. (1996). *Rights Across Borders.* Baltimore/London: The John Hopkins University Press.

John, P. (1998). *Analysing Public Policy.* London/New York: Continuum.

Jones, B. D., & Baumgartner, F. R. (2004). Representation and Agenda Setting. *Policy Studies Journal, 32*(1), 1–24.

Joppke, C. (1998). Why Liberal States Accept Unwanted Immigration. *World Politics, 50*(2), 266–293.

Jupp, J. (1993). Perspectives on the Politics of Immigration. In J. Jupp & M. Kabala (Eds.), *The Politics of Australian Immigration* (pp. 243–255). Canberra: Bureau of Immigration Research.

Kaye, R. (2001). An Analysis of Press Representation of Refugees and Asylum-Seekers in the United Kingdom in the 1990s. *Media and Migration: Constructions of Mobility and Difference, 8*, 53.

Kingdon, J. (1995). *Agendas, Alternatives, and Public Policies.* New York: HarperCollins.

Lahav, G. (1997). Ideological and Party Constraints on Immigration Attitudes in Europe. *Journal of Common Market Studies, 35*(3), 377–406.

Layton-Henry, Z. (1994). Britain: The Would-Be Zero Immigration Country. In *Immigration: A Global Perspective* (pp. 273–295). Stanford: Stanford University Press.

March, J. G., & Olsen, J. P. (2006). Elaborating the "New Institutionalism". In R. A. W. Rhodes, S. A. Binder, & B. A. Rockman (Eds.), *The Oxford Handbook of Political Institutions* (pp. 3–23). Oxford: Oxford University Press.

May, T. (2017, January 20). *Brexit Speech* [Online]. Available from: http://www.bbc.co.uk/news/uk-politics-38687842. Accessed 22 Jan 2017.

McLaren, L., & Johnson, M. (2007). Resources, Group Conflict and Symbols: Explaining Anti-immigration Hostility in Britain. *Political Studies, 55*(4), 709–732.

Menz, G. (2008). *The Political Economy of Managed Migration*. Oxford: Oxford University Press.

Meyers, E. (2000). Theories of International Immigration Policy—A Comparative Analysis. *International Migration Review, 34*(4), 1245–1282.

Mudde, C. (2007). *The Populist Radical Right in Europe*. Cambridge: Cambridge University Press.

Norris, P. (2005). *Radical Right: Voters and Parties in the Electoral Market*. Cambridge/New York: Cambridge University Press.

Odmalm, P. (2014). *The Party Politics of Immigration and the EU*. Basingstoke: Palgrave.

ONS (Office National Statistics). (2006). *International Migration* (No. 33). London: ONS.

ONS (Office National Statistics). (2012). *2011 Census, Key Statistics for Local Authorities in England and Wales*. Available from: http://www.ons.gov.uk/ons/rel/census/2011-census/key-statistics-for-local-authorities-in-england-and-wales/index.html. Accessed 11 Nov 2013.

ONS (Office National Statistics). (2014, November). *Migration Statistics Quarterly Report: November 2014*. London: ONS. https://www.ons.gov.uk/peoplepopulationandcommunity/populationandmigration/internationalmigration/bulletins/migrationstatisticsquarterlyreport/2015-06-30. Accessed on 20 Dec 2016.

Portes, J. (2016). Immigration—The Way Forward. In R. E. Baldwin (Ed.), *Brexit Beckons: Thinking Ahead by Leading Economists* (pp. 105–111). London: CEPR press.

Saggar, S. (2003). Immigration and the Politics of Public Opinion. *Political Quarterly, 74*(1), 178–194.

Sassen, S. (1996). *Losing Control?: Sovereignty in an Age of Globalization*. New York: Columbia University Press.

Schmidt, M. (1996). When Parties Matter: A Review of the Possibilities and Limits of Partisan Influence on Public Policy. *European Journal of Political Research, 30*(2), 155–183.

Soysal, Y. N. (1994). *Limits of Citizenship: Migrants and Postnational Membership in Europe*. Chicago: Chicago University Press.

Tichenor, D. J. (2002). *Dividing Lines: The Politics of Immigration Control in America*. Princeton: Princeton University Press.

Triadafilopoulos, T., & Zaslove, A. (2006). Influencing Migration Policy from Inside: Political Parties. In M. Giugni & F. Passy (Eds.), *Dialogues on Migration Policy* (pp. 171–193). Oxford: Lexignton Books.

van der Brug, W., D'Amato, G., Ruedin, D., & Berkhout, J. (2015). A Framework for Studying the Politicization of Immigration. In W. van der Brug, G. D'Amato, D. Ruedin, & J. Berkhout (Eds.), *The Politicisation of Migration* (pp. 1–17). Oxon: Routledge.

Wright, C. (2010). *Policy Legacies and the Politics of Labour Immigration Selection and Control* (Cambridge University PhD dissertation).

Wright, C. (2012). Policy Legacies, Visa Reform and the Resilience of Immigration Politics. *West European Politics, 35*(4), 726–755.

Zolberg, A. (1989). The Next Waves: Migration Theory for a Changing World. *International Migration Review, 23*(3), 403–430.

A Framework for Understanding Immigration Policy

2.1 Introduction

Unpacking the black box of immigration policymaking has long been a challenge for scholars. Since the 2000s, there has been a rapidly expanding literature that explores immigration policies of individual receiving states but there has been 'little debate on the relative merits of various schools of thought on the subject' (Meyers 2000, 1245). Scholars diverge as to what drives immigration policies in particular directions and which actors influence such processes. A particular challenge for researchers is to explain the gap between protectionist public preferences and the inclusionary policies that often emerge. Questions as to why states adopt expansive policies in the face of anti-migrant public sentiment and pressures for closure 'apparently endemic in democratic welfare states' (Boswell 2007, 75) have then long dominated the field. This raises questions as to who makes immigration policy, what the relationship is between political actors and policy outcomes (Schain 2008), and how the interactions among different interests, ideas and institutions mediate and shape policies.

Three approaches that explain government action and political elite preferences to immigration have come to dominate the field in political science: interest group, partisan politics and institutionalist approaches. The interest group or organised interests approach argues that non-governmental actors lobby and persuade governments to adopt specific immigration policies. Partisan politics places the onus on the role of political parties, their electoral strategy and their ideological disposition, as explanations for policy outcomes. In contrast, the institutional approach brings the state back in as an actor, and argues that it is the institutions of policymaking which determine policy. Whilst there is a degree of overlap among these perspectives (especially as each approach has several variants), the approaches fundamentally differ in terms of which sets of actors determine the direction of immigration policy: non-state actors, political parties and governmental actors respectively. In turn, the three approaches have different assumptions about how power is distributed across the state. This book draws on these approaches and examines the explanatory power each holds in the case study of Labour's managed migration regime.

The aim of this chapter is to provide an overview of the three main approaches adopted. The point is not to give an in-depth and exhaustive account of each approach, but rather to explain the logic and framework within which this research is situated and set out the research design and methodology of this study. This book is about explaining how rapid policy change occurs, but to do this we first need to consider what policy change actually means.

2.1.1 Defining Policy Change

Public policy scholars draw a broad distinction between normal change and atypical change. Streeck and Thelen (2005, 8) also make a comparable distinction between 'incremental' changes on the one hand and 'abrupt' changes on the other. Normal policy change 'involves relatively minor tinkering with policies and programs already in place through successive rounds of policymaking, which results in new policies being "layered" on top of existing ones' (Howlett et al. 2009, 202). Normal policy change is thus incremental and does not challenge existing policy paradigms. This being said, collectively these normal changes can affect the consistency of a policy regime. More exceptional and thus subject to scholarly research is atypical change. Atypical policy change involves 'changes in basic sets of policy ideas, institutions, interests and processes' (Howlett et al. 2009, 202). It is precisely because these major changes fundamentally conflict with the impediments of previous policy legacies that make such changes infrequent.

Similar to the distinction between normal policy change and atypical policy change, Peter Hall (1993) famously disaggregated policymaking into three orders of change in his influential study of economic policymaking in Britain in the 1970s–1980s. Hall suggests that policymaking 'involves three central variables: the overarching goals that guide policy in a particular field, the techniques or policy instruments used to attain these goals, and the precise settings of these instruments' (Hall 1993, 278). Building on these variables, Hall identifies three types of policy change:

> First and second order change can be seen as cases of "normal policymaking," namely of a process that adjusts policy without challenging the overall terms of a given policy paradigm…Third order change, by contrast, is likely to reflect a very different process, marked by the radical changes in the overarching terms of policy discourse associated with a "paradigm shift." If first and second order changes preserve the broad continuities usually found in patterns of policy, third order change is often a more disjunctive process associated with periodic discontinuities in policy (Ibid, 279).

First order changes refer to amendments to policy settings, such as adjusting tax rates. Second order changes denote alterations to the instruments of policymaking, such as the use of cash limits. Third order changes indicate shifts in the 'underlying assumptions and ultimate goals of policy itself', for example growth rather than unemployment (Freeman 2008, 375). While first and second order changes are largely technocratic affairs, third order changes can be a much 'broader affair subject to powerful influences from society and the political arena' (Hall 1993, 288). Of particular importance in Hall's framework is the way 'in which a paradigm serves to make sense of the world, to identify certain phenomena as problematic, and to suggest certain courses of action in response to them' (Freeman 2008, 375). Hall's differentiation proves the most useful for this research, as it serves to distinguish atypical or third order policy change from changes to the instruments of a policy regime. This book focuses on a third order change because Labour fundamentally transformed the objectives, framing, norms and political discourse of immigration policy. The book uses these terms throughout to refer to the overall shift in ideas and objectives of economic immigration policy, otherwise known as 'managed migration'. We now move to the three approaches adopted in this study.

2.2 Organised Interests: Non-governmental Actors and Policy Networks

One influential approach to policy change examines the role of organised interests; groups that seek to influence the public policy agenda and policy outcomes without competing for electoral office. Organised interests matter because they influence the policy agenda; they try to influence legislative and executive decisions, they often participate in the decisions concerning the implementation of policy and regularly implement the policies themselves (John 1998, 66). The organised interest approach in broad terms argues that immigration policy is a product of bargaining and compromise between government and interest groups. In turn, policy change is a result of interest groups pressuring government to change policy to meet their demands.

The approach stems from the pluralist tradition, which assumes that groups rather than individuals are crucial to understanding politics. In contrast to Marxist and elitist perspectives, pluralism assumes that in liberal societies there is a dispersal of power (Smith 1993, 36). Despite a myriad of critiques over the last century, the pluralist tradition remains strong and has reinvigorated itself over the past decade in a range of fields, including a notion of governance, the fundamental premise of which is that the central state is no longer the dominant force in determining public policy (Smith 1993, 31).

Yet pluralism has attracted much critique because it 'pays little attention the nature of the state and even less to state theory' (Smith 2006, 21). The state is usually seen as a monolithic, benign and neutral organisation, and this assumption means that pluralists have failed to examine how 'the state contains considerable power that often is not used for benign ends or distributed evenly between groups' (Smith 1993, 37). Taking the debate of power and the state as a starting point, Theodore Lowi (1964) suggested that both elitist and pluralist perspectives did not accurately describe power and public policymaking because 'both these schools mistake the resources of power for power itself' (Lowi 1964, 679). In response Lowi proposed a scheme to analyse public policymaking based on the proposition that relationships are determined by their expectations. Building on this proposition, Lowi argued that governmental policies condition expectation and therefore the type of policy at stake determines a political relationship. Consequently Lowi claimed that 'for every type of policy there is likely to be a distinctive type of political relationship' (Lowi 1964, 688).

Similarly to Lowi, James Q. Wilson (1980) challenged the assumption that 'federal politics were essentially legislative politics, dominated by political parties and carried out more or less in an arena of clearly competing interests' (Wilson 1980, viii). He contended that 'the ambitions of bureaucrats are reinforced by the electoral needs of congressmen and the private claims of interest groups' (Wilson 1980, viii). Against the notion that government officials are selfless, altruistic individuals, Wilson argued that officials respond to rewards much like businessmen and consumers (Wilson 1980, 361). Consequently, Wilson suggested that policy proposals are assessed in terms of their perceived costs and benefits. The political significance of costs and benefits arise out of their distributional impacts. In other words the distribution of consequences affects the incentive to form political organisations and engage in collective action (Wilson 1980, 366). These costs and benefits can be widely distributed or narrowly concentrated.

The American federal system, on which Lowi's and Wilson's research is based, offers multiple avenues for interest groups to lobby government, and thus arguably non-governmental influence is perhaps more prevalent than in majoritarian systems such as Britain. Nonetheless, since the 1990s, there has been a surge in non-governmental actors involved in policymaking in Britain, and this has led to an academic interest in how non-state actors form networks and shape public policy. Looking specifically at the role of policy networks in British policymaking, Rod Rhodes famously proposed a Policy Network Analysis model (1986, 1997). The underlying contention is that when the network structure of a policy sub-system alters, policy change may occur, such as when policy networks integrate new actors. According to Rhodes, the degree of conflict or consensus determines the nature of policy change. For example, we would expect policy change to be only incremental in a tightly knit policy community conditioned by value consensus. However, critics rightly contend that the network approach lacks explanatory value, or as Dowding puts it, is a 'metaphor rather than a model' (Dowding 1995). Change is ultimately explained with reference to the ideas that new actors bring into the network or exogenous changes that impact on the policy subsystem. In other words, 'they [network approaches] fail because the driving force of explanation, the independent variables, are not network characteristics per se but rather characteristics of components within the networks' (Dowding 1995, 137).

2.2.1 Policy Context: Organised Interests and Immigration Policy Change

Drawing explicitly on the work of Lowi and Wilson, Gary Freeman (1995, 2006) has developed the most convincing account of how organised interests influence policy outcomes. Freeman contends that the absence of well-organised anti-migrant groups drives policy in a specific direction.

His theory seeks to explain specifically why states adopt expansive immigration policies in the face of anti-migrant public discourse. Freeman and others (Cornelius et al. 2006) argue that there is an opinion-policy gap in immigration with policies often shaped 'out of public view and with little outsider interference' (Freeman 1995, 886). Therefore, they argue that immigration policy in most liberal democracies suffer from a permanent democratic deficit (Freeman et al. 2013, 2).

Contrarily some scholars claim that public opinion does play a role in immigration policy outcomes. Citrin and Sides (2008, 36) for example found that 'where leaders periodically must seek re-election to stay in power, public opinion does constrain what officials may safely contemplate'. Lahav (2004, 1158) similarly observed that public opinion matters in immigration policymaking because it may be:

> Politicized by elites who can then convert issues such as immigration or European integration and translate them on to the political agenda…public opinion is important in that it sets down the norms, or the 'rules of the game' by which elites must structure their discourse.

As Freeman et al. (2013) argue public opinion determines policy outcomes to the extent that it sets boundaries for elites. At times these boundaries allow for more generous numbers but in other times 'such as today, the boundaries are relatively narrow, thereby limiting policy options to the status quo or more restrictive measures' Freeman et al. 2013, 3). Whilst Freeman's model was certainly persuasive over 15 years ago, with the 'rise of anti-immigration parties and a more prominent discourse on immigration by many mainstream parties, the situation has changed considerably in recent years' (Morales et al. 2015, 4), and whether a democratic deficit remains in immigration policy across liberal states is debatable.

In terms of having influence over policy outcomes, Freeman (1995) tells us that pro-migrant lobby groups are better organised than anti-migrant groups. Political elites respond to such mobilised demands, which lead to immigration policies that are surprisingly expansive in light of public demands. Drawing on Wilson's and Lowi's frameworks, Freeman (2006) constructs a model that predicts how involved and influential interest groups will be in immigration policymaking depending on how distributive or concentrated the costs and benefits of particular policies are (Ibid, 229). According to Freeman, the benefits and costs of migration policies can be diffused or concentrated, resulting in four different modes of politics – client, majoritarian, interest group and entrepreneurial. In this model, if benefits are concentrated and costs are diffused, clientelistic politics will prevail, where small groups have a particular incentive to organise, and the wider public are indifferent or believe they will be unaffected by the resulting policy direction. Immigration policies are said to mirror the interests of those who have the most resources, and these tend to be those who stand to gain from expansive policies. According to Freeman, the principal beneficiaries of immigration are employers in labour-intensive industries and those dependent on an unskilled workforce, businesses that profit from population growth such as the construction sector and the families of those making up the immigration flows.

Georg Menz (2008) in his comparative study of labour migration policies has extended and refined Freeman's theory in terms of the positions interest groups take on labour migration. Menz suggested that employers would not simply push for more liberal policies, but rather the 'production system in which they [employers] are embedded conditions the quality and quantity of labour migration advocated' (Ibid, 8). In contrast to Menz, Alexander Caviedes (2010) argued that 'changes in the overall structure of the labour market have led to changes in employer preferences that differ to a large degree depending on sector' (Caviedes 2010, 2). Caviedes suggested that rather than an overarching model of capitalism determining employers' preferences, the real determinants of employers' preferences to immigration entail a 'sector-specific logic' (Caviedes 2010, 3).

Examining UK immigration policy specifically, Will Somerville (2007) and Somerville and Goodman (2010) explored the role of organised interests under the Labour governments of 1997–2010. Drawing on Rod Rhodes' (1986) distinction between issue networks and policy communities, Somerville and Goodman suggested that businesses and private sector interests were akin to a policy community underpinned by a high degree of consensus. The authors divided the organised public into five nodes comprising individual businesses and employers, legal associations, committees/lobbying groups, think tanks and the government. Somerville and Goodman (2010) argued that non-governmental actors comprising the policy network of labour migration had a substantive impact on economic immigration policy and that such actors were responsible for the policy change.

Conversely, Statham and Geddes (2006) found that immigration policymaking remained dominated by political elites in Britain. These elites act relatively autonomously and have their own political agenda (Statham and Geddes 2006, 266). Statham and Geddes examined the multi-organisational field and considered whether collective action influenced immigration policies towards expansion or restriction, with a focus on the asylum sub-policy field specifically. Their findings concluded that immigration policymaking in the British context was elite-driven and that Freeman strongly overstates the power of the organised public relative to national political elites (Statham and Geddes 2006, 266), suggesting that organised interests are not as powerful in Britain.

Another scholar who has examined the case under study here through a networks perspective deserves mention. Alex Balch (2009, 2010) adopted Haas's (1992) epistemic communities model to analyse the case of economic immigration policy change under Labour. Epistemic communities are networks of professionals with recognised 'expertise and competence in a particular domain' (Haas 1992, 3). This is comparable to John Kingdon's (1995) concept of the 'policy primeval soup' where 'policy entrepreneurs' including policy experts and non-government actors congregate in specific policy sub-systems to offer policy solutions to policy problems. The epistemic communities model argues that 'consensual knowledge'—which is defined as knowledge produced using certain scientific norms (Haas 1992)—crucially supplies policymakers with causal connections between ideas and policy application and this knowledge is more likely to be produced by epistemic communities. Balch's analysis focused on how experts and knowledge exerted influence on policy, rather than the influence of the business community or the third sector, but his analysis nonetheless found a determinative role for non-state actors. Balch argued that the policy shift was a result of 'a specific network of actors in the policy community, operating as an epistemic community, [which] dislodged the dominant restrictive policy paradigm or frame, and opened the debate around policy' (Balch 2009, 628). The principal determinant of policy change for Balch then was the network of 'experts' that provided new knowledge and thus ideas.

While the policy network approach, the epistemic communities model on which Balch draws, and the political economy models of Menz and Caviedes all differ in how and why non-state actors influence policy, all three models attribute a significant role to non-state actors or organised interests in policy change; indeed they place primacy on them.

Chapter 4 examines the role of organised interests in immigration policymaking under Labour using Freeman's model as a framework. The chapter pays particular attention to the ways in which non-governmental actors attempted to influence policy. If the organised interests approach has explanatory power in the British context of immigration policy change, we would expect to see lobbying from non-state actors in the late 1990s and early 2000s. In particular, one would expect employers and employer associations (especially those representing sectors experiencing labour market shortages) to have lobbied reasonably forcefully for more expansive economic immigration policies. In response, so the organised interest theory goes, the Labour governments responded to these mobilised and well-articulated demands to gain electoral support, whilst ignoring the inarticulate anti-migrant public.

2.3 Do Parties Matter: Political Parties and Party Ideology

An alternative explanation for policy change is associated with the party configuration of government. According to the 'parties-do-matter' hypothesis, the 'major determinant of variation in policy choices and policy outputs in constitutional democracies, is the party composition of government' (Schmidt 1996, 155). Stemming from an elitist perspective, those who adhere to the 'parties-do-matter' school of thought argue that political parties and the actors that comprise them have their own agenda and ultimately determine policy according to their partisan interests. Parties, in terms of party strategy and party ideologies, are thus said to matter in shaping and formulating policy. Without denying the importance of external constraining factors on political action such as socio-economic factors, 'advocates of the "politics matter" school of thought argue that there is a correlation between partisan variables and policy outputs' (Imbeau et al. 2001, 1). Proponents of the 'partisan hypothesis' assume then that there is a 'law-like tendency of partisan differences in public policy' where differences in public policy are 'significantly associated with—and, by inference, dependent upon—differences in the party composition of government' (Schmidt 1996, 156). In turn, 'a change in the party composition of government is associated with—and, by inference, causally related to—changes in policy choices and policy outputs' (Schmidt 1996, 156).

Political parties are distinct from interest groups as they act as the official representatives of the public, compete in elections for public office and 'unlike interest groups, parties actually control the government' (Burstein and Linton 2002, 385). The actors that comprise parties—political elites—are ultimately the ones who enact policy and are capable of fostering policy change. Political parties constitute a set of actors that influence policy by both translating public opinion into policies in exchange for support and at times acting as agents of change based on ideologies (Schmidt 1996, 155). Accordingly, policy outcomes are a product of public preferences (and thus electoral strategy) and/or a reflection of a party's ideology, although these are not mutually exclusive.

Commentators and scholars often talk of the 'end of ideology' (Bell 1962), claiming that party politics is now predominantly associated with the art of 'statecraft' (Bulpitt 1996), which is 'to do with winning elections and being able to govern effectively, particularly in the matters of "high politics"' (Kavangh et al. 2006, 78). The old battles between left and right are said to have 'declined in range and intensity, as has the divisions between social classes' (Kavanagh et al. 2006, 76), which has led to 'catch-all parties' (Kirchheimer 1966). As a result, scholars who study parties have tended to focus their attention on how parties ratify their policies in line with public preferences, on the calculation that parties are office seeking and thus vote-maximisers, and rarely pursue ideological purity over electoral expediency. Spatial approaches to party competition and party strategy (Downs 1957) have therefore dominated analyses of the relationships between parties and their policy proposals. It is in part through the party competition of office seeking that parties change policy in order to win public votes.

Party strategy cannot however explain why policies that are at odds with public preferences, such as expansive immigration policies in the face of protectionist public preferences, transpire. The essence of party success is not solely winning office in any case, but rather the 'achievement of a satisfactory trade-off between ideological introversion and electoral extroversion, between principles and power' (Bale 1999, 7). Because electoral calculation and thus party strategy cannot explain Labour's liberalising immigration regime, I focus on the ideas, philosophies and values of the governing party as an explanation for policy change, with the assumption that the ideas of the Labour Party were somewhat reflected public policy outcomes, including immigration. However, for 'ideas to have an impact they have to interact with other favourable conditions' (Kavangh et al. 2006, 78), such as the willingness of leading actors to be attracted by the ideas and push them, and circumstances or dissatisfaction with the existing set of policies.

Party ideologies 'provide politicians with a broad conceptual map of politics into which political events, current problems, electoral preferences and other parties' policies can all be fitted' (Budge 1994, 245). Ideologies thus provide political elites with a cognitive guide or blueprint for how to respond to social issues and problems, based on some underlying values, principles and crucially prescriptive ideas about how society should operate. The party's core identity is in other words an 'ideological resource pool which it can use to make claims over issues, and distinguish its stance from other parties' (Statham and Trenz 2013, 113). These are not binding frameworks however; ideologies rather supply parties with a way of

> indicating the broad area within which a particular party should take its position [which are] sufficiently ambiguous to permit some movement within the prescribed area, particularly if this simply involves the adjustment of policy priorities within the ideology itself rather than abandoning previously prescribed positions (Budge 1994, 246).

Many studies treat ideologies as 'reified objects based on static values, concerns and debates' (Bevir 2000, 279). Yet party ideologies are not fixed entities, they are rather a contingent set of ideas in which no value or identity has a fixed, central or defining place. Ideologies are, as Micheal Freeden (1996) puts it, 'morphological', constantly adapting and modifying according to the political context in which they are embedded. Ideologies consist of ideas and practices that people produce through their actions (Bevir 2000, 280), and because 'the evidence for political thinking is unobservable...its reflection must be sought in acts—expressions of writing, of doing' (Freeden 1990, 11), such as in party manifestos and political rhetoric. Because of the methodological limitations of determining the influence of party ideology, Bevir (2000, 282) argues that scholars should adopt a decentred approach to the study of ideology, and in doing so they,

> Should trace historical connections back through the immediate influences on the case we are explaining to the beliefs and practices that constitute the inheritance of the influences. By doing so, we would not define ideologies in terms of a given content – whether perennial or contingent and historicist – but, rather, pragmatically in relation to that which they explain.

However, it would be a simplification to suggest that ideology commands or dictates political elite action. As Szczerbiak and Taggart (2008, 257) argue, 'it is not possible to 'read off' a party's position from whatever ideological family it belongs to', in part because as above-mentioned parties are primarily office seeking but also because parties do not operate autonomously. Governing parties can and do face constraints from within the party and in terms of external competition from other parties. Parties seek to "own" issues of importance to win votes, and in doing so opposition parties 'help to cue, channel and even ramp' issues up, such as immigration (Bale 2008, 454). Therefore, governing parties face constraints from opposition parties who may coerce them into shifting or co-opting their rival's policies in a bid to maintain office. It is rare to see a government defeated by its own members, but nonetheless MPs can and do collectively rebel and in turn potentially block policy. Parties are rarely consensual homogenous organisations; they are rather prone to division, rebellion and factionalism. Thus, governing parties can also face internal constraints in implementing policy.

The constraints and laws of the political market suggest that party differences in public policy will be moderate, with radical differences being the exception rather than the rule (Imbeau et al. 2001). Despite the perhaps eroding influence of party ideologies in Western Europe, highly centralised states such as Britain are, 'in principle more amenable to partisan influences on public policy than states in which the government is constrained by counter-majoritarian powers such as federalism' (Schmidt 1996, 170). While the influence of party ideologies may well be deteriorating, the conduct of democratic politics is inconceivable without parties, and thus party ideology and party competition persist in playing some role in shaping public policy (Kavanagh et al. 2006, 350).

2.3.1 Political Parties and the Politics of Immigration

Immigration is an ideologically divisive issue for established parties of the left and right as it 'cuts across normal lines of political battle' (Lahav 1997, 382). There is no prescribed or evident approach to immigration policy for centrist parties (Hinnfors et al. 2012; Odmalm 2014), and some scholars have gone so far as to say that immigration represents an issue 'equated with the demise of ideology' (Lahav 1997, 382). Unlike most political issues where 'programmatic distinctions among parties serve to organise political debate and ultimately policy solutions' with immigration 'the process has not been so clear' (Lahav 1997, 382).

Kriesi et al. (2008) argued that the effects of globalisation and denationalisation have created a division between 'winners' and 'losers' in Western Europe. Winners are 'people who benefit from new opportunities resulting from globalisation, and whose life chances are enhanced' (Kriesi et al. 2008, 4). In contrast, the 'losers' 'are people whose life chances were traditionally protected by national boundaries. They perceive the weakening of these boundaries as a threat for their social status' (Kriesi et al. 2008, 4). Kriesi et al. (2008) argued that globalisation has led to a transformation of the cleavages that hitherto structured electoral competition. Through an emphasis on cultural issues such as immigration and European integration, Kriesi et al. (2008) contend that the traditional focus of the political debate—the economy—has been repurposed in this new political cleavage. The cultural dimension 'is now dominated by the issues of immigration, of cultural liberalism and European integration' (Lachat and Kriesi 2007, 4), 'which correspond to the new political and cultural forms of competition linked with globalization' (Lachat and Kriesi 2007, 5). Thus, according to Kriesi et al. (2008), as a result of globalisation immigration is now an issue equated with the new cultural dimension. The winners and losers of the globalisation thesis have never been more relevant, with commentators and scholars alike highlighting these divisions as an explanation for the Britain's surprising referendum result in June 2016. The losers of globalisation are de facto often referred to these days as the 'left behind', and the issue of whether these voices are marginalised is somewhat dominating debate at the time of writing (see Gest 2016).

Immigration poses an ideological problem for the centre right since part of its raison d'être is to defend the socio-economic and cultural status quo that immigration is seen to challenge (Bale 2008). The centre right "owns" policy issues akin to law and order and national security, which immigration is presumed to threaten. Because centre-right parties offer restrictive policies to minimise such threats, the centre right is in turn said to 'own' the issue of immigration (Green and Hobolt 2008). Centre-right parties are rooted in a relatively strong notion of national identity sketched around tradition and historical legacy (Baker et al. 2002, 402). Consequently, centre-right parties tend to politicise immigration by focusing on questions of national identity and the apparent challenge that immigrants pose to this (Schain 2008, 467).

Yet, fundamentally, immigration prompts tensions between the 'identity right', the 'paternalistic right' and the 'business right' for centre-right parties (Bale 2008). These tensions are particularly pertinent in the British case, where 'the Tories have been deeply divided between English nationalists and free trade internationalists' (McLean 2001, 144), and this problem is especially acute for the neoliberal New Right, given that 'free trade would logically entail free movement of workers' (Smith 2008, 420). Again, these tensions are extant in today's political context, where arguably the referendum on Britain's membership of the EU partly came about precisely because of the intra-party divisions in the Conservative Party over European integration.

Likewise, immigration is a divisive issue for the centre left. On the one hand, parties of the centre left tend to support immigration because 'they defend a universalist position of solidarity with often marginalized and oppressed communities' (Keith and McGowan 2014) and because in turn migrants are seen as a source of working class support. Subsequently one would expect left parties to treat 'vulnerable and exposed groups such as refugees and immigrants with 'open, generous and solidaristic means', which 'sit comfortably with social democratic foundations' (Hinnfors et al. 2012, 589). On the other hand, parties of the left have often supported restrictive and exclusionary policies for two reasons. First, employers have in the past used migrant labour to suppress wages and break strike movements. Second, the reaction to immigrants by native working class voters has often made left parties electorally vulnerable (Schain 2008). Furthermore, as Odmalm and Bale (2015, 5) note, in the end "uncontrolled" immigration could potentially create 'a new—ethnic—underclass and accordingly split the indigenous vote'. Immigration confronts centre-left parties with two contradicting ideological pulls then 'international solidarity versus welfare state/labour market protectionism' (Odmalm 2011, 1071). And like the centre right, the far right poses a problem for the centre left, as the populist far right have mobilised immigration and targeted the centre left's traditionally core constituents (Ford and Goodwin 2010). In response, social democratic parties can hold their position, adopt a tougher rhetoric on immigration or attempt to defuse the saliency of the issue by emphasising other matters (Bale et al. 2010).

In consideration of the contradicting ideological pulls immigration presents (Odmalm 2014), centrist parties across Western Europe have unsurprisingly tended to 'downplay the issue, because of the cross-cutting cleavages that affect their core constituencies' (Perlmutter 1996, 377).

Immigration not only divides social constituencies for both the centre left and the centre right but also 'is as likely to divide one's own party as one's opposition' (Perlmutter 1996, 377). This defusing of immigration and race politics has traditionally led to party convergence or a 'hidden consensus' made in order 'to include mass parties' interest in not alienating important constituencies and internal factions, precluding the rise of racial tensions, and maintaining relatively liberal positions despite negative perceptions regarding immigration held by the majority of the population' (Triadafilopoulos and Zaslove 2006, 171). Yet ignoring the issue or in other words the defuse strategy (Bale et al. 2010) can lead to the populist radical right hijacking the political space and offering the electorate restrictive policy solutions that chime with public preferences. Ultimately the ideal strategy is to achieve a balance 'between emphasising policy strengths and engaging with immigration so it does not bring the conflicting ideological "pulls" to the fore' (Odmalm 2011, 1084).

Despite political attempts to defuse immigration, partisan ideologies still shape elite preferences and policy outputs to some extent, yet party ideology has been overlooked in the literature on parties and immigration (Hinnfors et al. 2012, 588). Givens and Luedkte (2005) amongst others (Money 1999; Ireland 2004) found that partisanship plays a role in integration policy, but that 'Left parties are no less restrictive than Right parties on immigration control' (Ireland 2004, 16). Money (1999) suggests that parties on the left are more likely to pass expansive integration policies, as this will extend immigrant rights including voting rights. From this perspective, expanding integration policy is a strategy to mobilise potential voters. However, parties of the centre left cannot afford to be liberal in terms of immigration control policies, as short-term public backlash could offset any future gains from an expanded immigrant electorate. Money's account suggests that party affiliation only plays a role in terms of electoral strategy rather than fundamental ideological differences.

Conversely, other scholars have suggested that partisan ideology does play a role in shaping elite preferences to immigration policy. For example, Ireland (2004) in his case study of German elite preferences concludes that a portion of the centre-left party was ideologically committed to cultural pluralism. Lahav (1997, 401) in her study of MEP attitudes to immigration also shows that 'although the immigration issue has appeared to obscure ideological/party differences, party affiliation persists in differentiating attitudes towards immigration'. She goes on to suggest that 'the left-right constructs order preferences in a way that partisans of the left are more likely than their colleagues on the right to endeavour to amend social inequalities and to extend immigrant rights' (Lahav 1997, 401). According to Lahav, party ideology does matter in shaping political elites' preferences to immigration, to the extent that it 'continues to shape the dialogue on immigration intake, and educate the public to the various policy dilemmas and goals' (Lahav 1997, 401). Hinnfors et al. (2012) similarly found that party ideology rather than electoral strategy explains the Swedish Social Democratic Party's response to immigration. They argue that 'room must be made in explanatory narratives for the role of party ideology' (Hinnfors et al. 2012, 586–7). If political elites' preferences are shaped by party ideologies, given that political elites direct and enact policy, we would expect these ideas to be at least somewhat reflected in policy.

The contradictory ideological pulls that immigration presents combined with the saliency of the issue in the first instance irreducibly pushes parties of the left or right to pursue or maintain restrictive immigration policies. With a public and mass media hostile to immigration, the rising politicisation of immigration and the ideological challenges established parties must confront, it would always be easier and electorally logical for a centre-left party to maintain the status quo of restrictive immigration policies. After all, 'in no Western European country can politicians or political parties gain votes by favouring new immigration' (Lahav 1997, 382). Being confronted with these issues makes the expansionary reforms under Labour even more puzzling. However, a lens that sees parties as simply vote-maximisers has dominated the literature on parties and immigration, and thus analysis has focused on electoral strategies (Bale et al. 2010), party competition (Odmlam 2011) and the threat from the emerging populist radical right (Mudde 2013). Lacking in these accounts is how party ideology can be an explanatory factor for why parties formulate their immigration policies in the way they do, especially for social democratic centre-left parties that have been somewhat overlooked in the literature (see Hinnfors et al. 2012 and Odmalm 2014 for exceptions). This is perplexing and especially pertinent to this case study, as New Labour's ideology had a discernible impact on many other policy areas, including the economy (Watson and Hay 2003), education (Cole 1998) and health (Harrison 2002).

If 'parties-do-matter' (Schmidt 1996), we would expect the governing party ideology to exert an influence on the immigration policy preferences of the leading elite. Proving that party leaders have been affected by any one factor is a methodological challenge, but by adopting a decentred approach to the study of ideology, the historical tenets and evolution of the Labour Party can be mapped on to the shifts in the Party's immigration policy. We would also anticipate that party competition to 'own' the issue, and conceivably internal divisions within the governing party, possibly constrained government action and policy implementation. Chapter 5 examines whether the governing party's ideology was reflected in immigration policy. The chapter also examines the political context by considering the external (opposition parties) and internal (division within the party) constraints placed on the party in terms of implementing immigration policy.

2.4 Historical Institutionalism: Institutional Change, Policy Framings and Critical Junctures

In contrast to those who focus on party political interactions, a third perspective looks to the state itself and the institutions that comprise it to explain government decisions on immigration policy. The historical institutional approach 'brings the state back in' by focusing on how administrations and bureaucracies shape policy. Historical institutionalism stresses the way in which the actions of individuals exist within the context of rules and norms and through the legacies of previous decision-making (Lowndes 2002).

The new institutionalist school—which historical institutionalism stems from—claims that political institutions can be autonomous and that it is these apolitical institutions that make policy according to the interests of the state. The new institutionalist approach is diverse and includes several variants that differ according to the degree of autonomy attributed to the state as well as conflicting understandings of structure and agency. Scholars also diverge on whether the state is monolithic or whether various bureaucratic agencies and institutions pursue their own interests and agendas. Nonetheless, all variants of new institutionalism contend that institutions are a set of relatively enduring rules and organised practices, which are somewhat resilient to idiosyncratic preferences and the external environment (March and Olsen 2006, 3). Institutions are said to both constrain and condition the behaviour of political actors (Béland 2005; Steinmo et al. 1992) and therefore understanding institutions and institutional adaptation is fundamental to explaining policymaking and policy change.

In contrast to other strands of new institutionalism—such as rational choice—historical institutionalists focus more explicitly on the state, its institutional development, and how such historically embedded institutional developments constrain individual behaviour (Schmidt 2006). Historical intuitionalism emphasises how historical legacies structure and constrain political outcomes, a concept known as path dependency. It stresses the importance of sequences in development, timing of events and phases of political change and is thus particularly attentive to processes over time.

Given the emphasis on structures, one of main problems for the historical institutionalist is explaining human agency. In response, historical institutionalists have divided into those on the one hand who adhere to a 'calculus approach' (Hall and Taylor 1996, 940; Thelen 2004; Pierson 2004), which puts historical institutionalists closer to rational choice institutionalists, and on the other hand those who have adopted a 'culture' approach, which puts them closer to sociological institutionalists (Schmidt 2006, 106; Berman 1998; Blyth 2002). Many who have adopted a cultural understanding of human agency within the historical institutionalism school have focused on the role of ideas, 'tipping' this strand into discursive institutionalism, although 'the dividing line is fuzzy' (Schmidt 2006, 111). What characterises this strand of historical institutionalism is 'the focus on ideas as explanatory of change, often with a demonstration that such ideas do not fit predictable "rationalist" interests, are underdetermined by structural factors, and/or represent a break with historical paths' (Schmidt 2006, 111). Thus, when it comes to explanations of human agency, historical institutionalists have split between rationalist and constructivist understandings of institutions.

Ontologically this appears contradictory; for how can one approach contain both rationalist and constructivist assumptions about human agency. Hay and Wincott (1998, 954) argue that the social ontology of historical institutionalism means that such institutionalists should reject the view of the rational actor on which the calculus approach is premised, because actors cannot be assumed to have a fixed (and immutable) preference set, to be blessed with extensive (often perfect) information or to be self-serving utility maximisers. Chapter 6, which examines the institutional context of policymaking, adopts the 'culture' understanding of human agency within the historical institutionalist school.

The 'cultural' version of historical institutionalism starts from the assumption that the different components of the state—agencies, bureaucracies and departments—are built on ideational foundations; that is, the initial construction of institutions are inaugurated by ideas. These institutions develop ideas and framings on the policy areas within their remit in an autonomous manner. This is because whilst institutions condition and constrain action, 'they are themselves also the outcome (conscious or unintended) of deliberate political strategies of political conflict and choice' (Thelen and Steinmo 1992, 10). Thus whilst institutions are structural in that they condition behaviour, they are also constructed in that they are ultimately the products of human agency.

The historical institutionalist interpretation of ideas, at least within the 'culture' strand, is that they function as both mobilising forces for collective actors who want to create or change institutions and as the standards of evaluation for policy outcomes in existing institutions (Sanders 2006, 42). Through processes of normalisation and socialisation, certain ideas, objectives and framings become embedded in these institutions that serve as cognitive filters through which actors (namely civil servants in this context) come to interpret their environment. Ideas then become codified into institutional practice and 'take on a life of their own' (Berman 1998, 18) to the extent that they become resilient to change. Crucially, historical institutionalists are concerned with how, under certain conditions, such institutionalised ideas and paradigms are contested, challenged and replaced (Hay 2006, 65).

The 'cultural' historical institutionalist assumes that actors' desires, preferences and motivations are not a reflection of material or social circumstances, but are ideational and reflect a normative orientation (Hay 2006, 63). This assumption is grounded in the proposition that 'reality and actors' perceptions of it are not synonymous' (Berman 1998, 30); thus 'interests are neither given nor can they be inferred from the economic environment' (Berman 1998, 30). According to historical institutionalists who adhere to a constructivist understanding of institutions, reality is often too complex and/or ambiguous to be handled directly, and this leads political actors to rely on cognitive shortcuts to make sense of the world (Berman 1998, 31; Hay 2006). One such cognitive shortcut is the way in which over time, ideas and norms collectively form a policy frame that defines in a given policy sub-system principles of action, as well as methodological prescriptions and practices for actors subscribing to the same frame (Surel 2000, 496). In turn, policy frames can become fixed and institutionalised amongst actors:

Their [political actors] perceptions about what is feasible, legitimate, possible, and desirable are shaped both by the institutional environment in which they find themselves and by the existing policy paradigms and world-views. It is through such cognitive filters that strategic conduct is conceptualized and ultimately assessed (Hay 2006, 65).

In terms of explaining change, those who adopt a 'cultural' understanding of institutions contend that ideas can prompt a new policy framing, and in turn policy change, when the normal policy frame is seen as inadequate, otherwise known as 'incremental adjustment' (Pierson 2004). The historical institutionalist claims that new ideas and framings can also be mobilised through processes of 'policy learning' (Hall 1993; Sabatier 1988), which 'examines the reframing of policy issues with the possible reframing of the associated institutions' (Peters 2012, 79). Policy learning can be prompted by conflicts between alternative visions of policy or a deeper conflict over policy ideas (Peters et al. 2005). In this sense 'powering' and 'puzzling' are intertwined and 'the competition for power can itself be a vehicle for social learning' (Hall 1993, 289). However, these ideas will only be viable alternatives if influential political actors advocate them, for ideas remain just that without agents mobilising them. These actors are referred to as 'policy entrepreneurs' (Sabatier 1988; Kingdon 1995), 'agenda-setters' (Richards and Smith 1997) or 'carriers' (Berman 1998). For an idea to gain currency, it must be associated with political actors, and the more powerful these actors are, the more likely an idea will become accepted and embedded. The historical institutionalist does not then contend that ideas are free-floating abstractions that can occur at any time. Rather ideas as a causal force are dependent on institutional legitimisation by powerful actors.

Most institutionalist accounts point to crisis or exogenous factors, such as socio-economic change, as events that render the previous policy frame inadequate to the new set of circumstances. However, John Kingdon (1995) in his work on agenda setting demonstrated that a culmination of favourable dynamics could allow the opening of a political window, which is 'an opportunity for advocates of proposals to push their pet solutions, or to push attention to their special problems' (Kingdon 1995, 173). Kingdon suggests that the power of ideas in changing a policy frame is not confined to moments of crisis, but can be a result of a favourable set of circumstances that open a window of opportunity for actors to promote a new policy frame.

Another mechanism of policy change can be endogenous, internal processes of institutional change. One such process Kathleen Thelen (2004, 36) labels 'institutional conversion' where, 'the adoption of new goals or the incorporation of new groups into the coalitions on which institutions are founded can drive a change in the functions these institutions serve or the role they perform'. When new actors are integrated into an institution this can lead to additional elements being layered on an institution and in turn redirect the objectives (Thelen 2004, 293), such as when a previously dormant or inactive institution, for example a government department, becomes newly active in a given policymaking process (Thelen and Steinmo 1992, 17). In this way institutional change is not triggered when some external environmental factor changes; rather institutional change can occur, 'precisely when problems of role interpretation and enforcement open up spaces for actors to implement existing rules in new ways' (Mahoney and Thelen 2010, 4). Accordingly, institutional change can be internal, evolutionary yet still transformative. Whilst the focus of this study is on policy change as opposed to institutional change, institutional shifts impact on the norms, conventions and practices of public policymaking, and such shifts therefore have the capacity to cause policy change.

The power of institutions is in their constraining capacity, as institutional rules are relatively enduring and resilient to idiosyncratic preferences and expectations (March and Olsen 2006, 3), making them acquire a highly "layered" quality (Pierson and Skcopol 2002, 14). In turn, institutions tend to generate policy reproduction and stability. However, radical policy change is rare but still possible. These moments of radical transformation are described as critical junctures in the historical institutionalism literature, which are events or brief moments in which 'structural (economic, cultural, ideological, organizational) influences on political action are significantly relaxed for a relatively short period' (Capoccia and Kelemen 2007, 343). Critical junctures are then moments in time that allow for policy change, analogous to Kingdon's (1995) 'window of opportunity'.

Change is most dramatically prompted by an exogenous 'shock' to an otherwise settled institutional environment, which drives a revaluation of institutional practices. Major events such as wars or civil conflicts or economic crises are the most obvious examples. Alternatively, we can interpret a critical juncture as a moment where multiple variables internally change at the same time, which in turn creates a space for political change. Crucially, a critical juncture is usually caused by a combination of several independent variables that interact to produce a relatively short window of opportunity, in which political actors can break away from 'business as usual' (Schwartz 2001, 13). As a result, interdependency of variables becomes the key to explaining policy change: 'These contingencies matter because it is only during a critical juncture that actors are truly free to choose among competing alternatives' (Schwartz 2001, 13). What becomes essential in explaining policy change are the temporal conditions and the conjectures of a variety of internal political forces. Individually these are not capable of generating change, but when taken together the effect can be radical. Paradigmatic shifts therefore require a combination of time, circumstances and powerful actors (Richards and Smith 1997, 76).

Yet while critical junctures represent marked points at which new 'paths' are chosen, critical junctures should nevertheless be understood, indeed can only be understood, in relation to previous 'paths' (Consterdine and Hampshire 2014; Pierson 2004). The concept of path dependency, which is at the heart of historical institutionalist analysis, does not simply encapsulate the banal truth that 'history matters'. Rather it conveys how feedback mechanisms affiliated with earlier policy choices can constrain future options because deviating from such paths will have high political and/or institutional costs (Levi 1997, 27), hence why an atypical or third order policy change is so exceptional. A change is only a critical juncture then if there are visible institutional feedback effects that lay a new path for policy.

Path dependency is essentially a term used to signify how 'feedback mechanisms affiliated with earlier policy choices can constrain future options because deviating from such paths will have high political and/or institutional costs' (Levi 1997, 27). The key assumption of feedback effect theory then is that public policies reconfigure the rules of the game, 'influencing the allocation of economic and political resources, modifying the costs and benefits associated with alternative political strategies and consequently altering ensuing political development' (Pierson 1993, 596). Whilst research usually focuses on how politics effects policy or, in other words, policy is the dependent variable, feedback research asks how new policies create new politics (Pierson 1996). There are a number of ways a policy legacy can exert a feedback effect, including structural feedback effects on policymaking, such as interest group formation feedbacks or state building, and mass feedback effects where policies create feedback effects on publics and politics. Previous policies can also create symbolic feedback effects, where ideas have become codified and changed policy framings and norms. The key mechanism that generates a feedback effect is an increasing return 'which could also be described as self-reinforcing or positive feedback processes' (Pierson 2000, 251). Increasing return as a mechanism refers to the 'probability of further steps along the same path increases with each move down the path—this is because the relative benefits of the current activity compared with other alternatives increases over time' (Pierson 2000, 252), thus generating a self-reinforcement of a positive feedback effect.

2.4.1 Institutions and Immigration Policy

Given that 'the entire analytical framework of historical institutionalism appears premised upon the enduring effects of institutional and policy choices made at the initiation of a structure' (Peters 2012, 77), the approach is much better suited to explain policy stability. Unsurprisingly then relatively few scholars have employed this approach to explain immigration policy change (Consterdine and Hampshire 2014). The exception to this has been Virginie Guiraudon (2003) who employed Kingdon's (1995) concept of a window of opportunity to explain the Europeanisation of immigration policy. She argued that the establishment of EU antidiscrimination legislation transpired partly as a result of favourable timing, or a window of opportunity, for a particular group (Starting Line Group) that were able to informally set the agenda at a critical meeting (the 1996 Intergovernmental Conference) (Guiraudon 2003, 274).

Many scholars have however employed a historical institutionalist approach to explain trajectories of immigration policy development and have found path-dependent effects on immigration policymaking (Hansen 2000; Tichenor 2002; Wright 2012). Moreover, a number of scholars have analysed immigration policy with a disaggregated lens of the state, and found that the different institutions of immigration policymaking have divergent and often contradictory agendas on immigration policy, particularly between central government departments and the agencies that implement such policies (Wright 2010; Eule 2014; van der Leun 2006). Others have also engaged with the concepts of new institutionalism to explain comparatively why certain ideas and framings do or do not become embedded in the institutions of immigration policymaking. For example, Hansen and King (2001) examined how ideas stemming from the eugenics tradition translated into immigration policy outcomes in

Britain and the USA. They found that the idea failed to materialise in Britain due to timing and lack of institutional positioning by its proponents, thus reinforcing the interdependency between ideas and interests.

Other scholars have similarly analysed the role of ideas and policy framings in immigration policy (Bleich 2003). The supranational EU level for example frames immigration policy in either 'realist' terms as an issue of internal security or a 'liberal' frame of human rights (Lavanex 2001). Adrian Favell (1998) examined the role of ideas and framings in integration policy in France and Britain that formed what he called public philosophies. He found that in France the dominant policy framework addressed the country's ethnic dilemmas in terms of republican ideas. In contrast, Britain addressed these problems in terms of race relations and multiculturalism. Favell argued that these frameworks rest on different philosophies 'based on contrasting understandings of core concepts such as citizenship, nationality, pluralism, autonomy, equality, public order and tolerance' (Favell 1998, 2). These philosophies, he claimed, hold irrespective of the party composition of government and persevere over time.

Chapter 7 adopts a historical institutionalist lens to the question of policy change. Taking a disaggregated view of the state, the chapter examines the administrative context in detail, analysing the role of different government departments in immigration policymaking, their institutionalised policy framings of immigration, and assessing whether institutional changes to policymaking practices had an influence on policy outcomes. If the historical institutionalist approach has explanatory power in the case, we would expect the different institutions of immigration policymaking to have developed habitual modes of policymaking and framings of policy issues. We would predict that the sequencing, timing of policymaking and endogenous institutional shifts may have had an impact on the degree of policy change. The concluding chapter considers complex causality and reflects on whether the case constitutes a critical juncture.

2.5 Methodology and Research Design

The purpose of using three approaches or a 'multiple lenses approach' (Lewis and Grime 1999) to the question of policy change is to utilise different understandings of how and why policy change transpires. A multiple lenses approach is apt at revealing 'seemingly disparate, but interdependence facets of complex phenomena' (Lewis and Kelemen 2002, 258). Epistemologically the research is situated in the interpretivist tradition, albeit in a 'soft' understanding, because using a multiple lenses approach assumes that alternative explanations are possible and can work to complement each other to enhance our understanding of the social world (Lewis and Grimes 1999). A soft interpretative approach accepts that the different explanations of policy change are essentially different interpretations of the same phenomenon. Such an epistemology allows us to both 'respect opposing approaches and juxtapose the partial understandings they inspire' (Lewsi and Kelemen 2002, 258). This enables us to supply different narratives of policy change and evaluate which, if any, offers the best explanation based on the empirical data.

Rather than undertaking competitive testing of abstractly generated theoretical assumptions, which as Falkner (2002, 4) says 'easily turns the world into a black-and-white-only picture where manifold shades of grey, not to speak of all other colours, go unnoticed', this research adopted an explorative process-tracing design that was amenable to recognising unexpected dynamics and diverse causal effects. Whilst process tracing stems from the positivist tradition, this method can nonetheless contribute to the interpretivist case study approach (Checkel 2006) as it helps to uncover 'directly and indirectly what actors want, know and compute' (Vennesson 2008, 233) and is well suited to unravelling complex causation (George and Bennett 2005).

Since the objective of the research is to trace processes, decision-making and understand policymaking in a specific setting (Grix 2001, 33), qualitative research was better suited to gather data that can be subjected to interpretation and analysis. The research methods employed were archival analysis (from the National Archives and the People's History Museum), elite interviews and the analysis of grey literature including government documents such as white papers, green papers, official minutes, Hansard transcripts, consultation papers, press releases and speeches. Fifty interviews were conducted between June 2011 and May 2015 with a range of actors including former and current Labour politicians, former special advisors, trade unions, employer associations and sector specific interest groups, NGOs, think tanks and pressure groups. I also conducted interviews with civil servants from the Home Office, UK Visas and Immigration, the Treasury, the Department for Business, Innovation and Skills, and the Foreign and Commonwealth Office. To protect sources names of interviewees are anonymous throughout the book wherever possible.

The three approaches suggest a range of possible reasons why policy change occurs and, in particular, why states adopt expansive immigration policies. Whilst there is a degree of overlap between these approaches, each assumes that a different set of actors ultimately drives policy and policy change, and they therefore adopt divergent understandings of how power is distributed across the state. After providing contextual background of the case, the subsequent chapters examine whether the empirical data 'fit' with any of the approaches outlined. We now turn to Britain's post-war history of immigration to understand the context and degree of policy change under the New Labour administrations.

References

Baker, D., Gamble, A., & Seawright, S. (2002). Sovereign Nations and Global Markets: Modern British Conservatism and Hyperglobalism. *British Journal of Politics and International Relations, 4*(3), 399–428.

Balch, A. (2009). Labour and Epistemic Communities: The Case of "Managed Migration" in the UK. *The British Journal of Politics and International Relations, 11*(4), 613–633.

Balch, A. (2010). *Managing Labour Migration in Europe: Ideas, Knowledge and Policy Change*. Manchester: Manchester University Press.

Bale, T. (1999). *Sacred Cows and Common Sense: The Symbolic Statecraft and Political Culture of the British Labour Party*. Aldershot: Ashgate Publishing.

Bale, T. (2008). Turning Round the Telescope. Centre-right Parties and Immigration and Integration Policy in Europe. *Journal of European Public Policy, 15*(3), 315–330.

Bale, T., Green-Pedersen, C., Krouwel, A., Luther, K. R., & Sitter, N. (2010). If You Can't Beat Them, Join Them? Explaining Social Democratic Responses to the Challenge from the Populist Radical Right in Western Europe. *Political Studies, 58*(3), 410–426.

Béland, D. (2005). Ideas and Social Policy: An Institutionalist Perspective. *Social Policy and Administrative, 39*(1), 1–18.

Bell, D. (1962). *The End of Ideology: On the Exhaustion of Political Ideas in the Fifties: With "The Resumption of History in the New Century"*. Harvard: Harvard University Press.

Berman, S. (1998). *The Social Democratic Moment: Ideas and Politics in the Making of Interwar Europe*. Cambridge: Harvard University Press.

Bevir, M. (2000). New Labour: A Study in Ideology. *British Journal of Politics and International Relations, 2*(3), 277–301.

Bleich, E. (2003). *Race Politics in Britain and France: Ideas and Policymaking Since the 1960s*. Cambridge: Cambridge University Press.

Blyth, M. (2002). *Great Transformations: Economic Ideas and Institutional Change in the Twentieth Century*. Cambridge: Cambridge University Press.

Boswell, C. (2007). Theorizing Migration Policy: Is There a Third Way? *International Migration Review, 41*(1), 75–100.

Budge, I. (1994). A New Spatial Theory of Party Competition: Uncertainty, Ideology and Policy Equilibrium Viewed Comparatively and Tentatively. *British Journal of Political Science, 24*(4), 443–467.

Bulpitt, J. (1996). Historical Politics: Leaders, Statecraft and Regime in Britain at the Accession of Elizabeth II. *Contemporary Political Studies, 6*(2), 1093–1106.

Burnstein, P., & Linton, A. (2002). The Impact of Political Parties, Interest Groups and Social Movement Organizations on Public Policy: Some Recent Evidence and Theoretical Concerns. *Social Forces, 81*(2), 341–408.

Capoccia, G., & Keleman, D. R. (2007). The Study of Critical Junctures: Theory, Narrative, and Counterfactuals in Historical Institutionalism. *World Politics, 59*(3), 341–369.

Caviedes, A. A. (2010). *Prying Open Fortress Europe: The Turn to Sectoral Labor Migration*. Lanham: Lexington books.

Checkel, J. T. (2006). Tracing Causal Mechanisms. *International Studies Review, 8*(2), 362–370.

Citrin, J., & Sides, J. (2008). Immigration and the Imagined Community in Europe and the United States. *Political Studies, 56*(1), 33–56.

Cole, M. (1998). Globalisation, Modernisation and Competitiveness: A Critique of the New Labour Project in Education. *International Studies in Sociology of Education, 8*(3), 315–333.

Consterdine, E., & Hampshire, J. (2014, forthcoming). Immigration Policy Under New Labour: Exploring a Critical Juncture. *British Politics, 9*(3), 275–296.

Cornelius, W., Tsuda, T., Martin, P., & Hollifield, J. (2006). *Controlling Immigration: A Global Perspective*. Stanford: Stanford University Press.

Dowding, K. (1995). Model or Metaphor? A Critical Review of the Policy Network Approach. *Political Studies, 43*(1), 136–158.

Downs, A. (1957). *An Economic Theory of Democracy*. New York: Harper.

Eule, T. (2014). *Inside Immigration Law: Migration Management and Policy Application in Germany*. Surrey: Ashgate.

Falkner, G. (2002). EU Treaty Reform as a Three-Level Process. *Journal of European Public Policy, 9*(1), 1–11.

Favell, A. (1998). *Philosophies of Integration: Immigration and the Idea of Citizenship in France and Britain*. Basingstoke: Palgrave Macmillan.

Ford, R., & Goodwin, M. (2010). Angry White Men: Individual and Contextual Predictors of Support for the British National Party. *Political Studies, 58*(1), 1–25.

Freeden, M. (1990). The Stranger at the Feast: Ideology and Public Policy in Twentieth Century Britain. *Twentieth Century British History, 1*(1), 9–34.

Freeden, M. (1996). *Ideologies and Political Theory: A Conceptual Approach*. Oxford: Clarendon Press.

Freeman, G. (1995). Modes of Immigration Politics in Liberal Democratic States. *International Migration Review, 29*(4), 881–902.

Freeman, G. (2006). National Models, Policy Types, and the Politics of Immigration in Liberal Democracies. *West European Politics, 29*(2), 227–247.

Freeman, R. (2008). Learning in Public Policy. In M. Moran, M. Rein, & R. E. Goodin (Eds.), *The Oxford Handbook of Public Policy* (pp. 367–389). Oxford: Oxford University Press.

Freeman, G., Hansen, R., & Leal, D. L. (2013). Introduction: Immigration and Public Opinion. In G. Freeman, R. Hansen, & D. L. Leal (Eds.), *Immigration and Public Opinion in Liberal Democracies* (pp. 1–21). New York: Routledge.

George, A. L., & Bennett, A. (2005). *Case Studies and Theory Development in the Social Sciences.* Cambridge: MIT Press.

Gest, J. (2016). *The New Minority: White Working Class Politics and Marginality.* Oxford: Oxford University Press.

Givens, T., & Luedtke, A. (2005). European Immigration Policies in Comparative Perspective: Issue Salience, Partisanship and Immigrant Rights. *Comparative European Politics, 3*(1), 1–22.

Green, J., & Hobolt, S. (2008). Owning the Issue Agenda: Party Strategies and Vote Choices in British Elections. *Electoral Studies, 27*(3), 460–476.

Grix, J. (2001). *Demystifying Postgraduate Research: From MA to PhD.* Birmingham: University of Birmingham Press.

Guiraudon, V. (2003). The Constitution of a European Immigration Policy Domain: A Political Sociology Approach. *Journal of European Public Policy, 10*(2), 263–282.

Haas, P. (1992). Introduction: Epistemic Communities and International Policy Coordination. *International Organization, 46*(1), 1–35.

Hall, P. (1993). Policy Paradigms, Social Learning, and the State: The Case of Economic Policymaking in Britain. *Comparative Politics, 25*(3), 275–296.

Hall, P. A., & Taylor, R. C. (1996). Political Science and the Three New Institutionalisms. *Political Studies, 44*(5), 936–957.

Hansen, R. (2000). *Citizenship and Immigration in Post-war Britain.* Oxford: Oxford University Press.

Hansen, R., & King, D. (2001). Eugenic Ideas, Political Interests, and Policy Variance: Immigration and Sterilization Policy in Britain and the U.S. *World Politics, 53*(2), 237–263.

Harrison, S. (2002). New Labour, Modernisation and the Medical Labour Process. *Journal of Social Policy, 31*(3), 465–485.

Hay, C. (2006). Constructivist Institutionalism. In R. A. W. Rhodes, S. A. Binder, & B. A. Rockman (Eds.), *The Oxford Handbook of Political Institutions* (pp. 56–75). Oxford: Oxford University Press.

Hay, C., & Wincott, D. (1998). Structure, Agency and Historical Institutionalism. *Political Studies, 46*(5), 951–957.

Hinnfors, J., Spehar, A., & Bucken-Knapp, G. (2012). The Missing Factor: Why Social Democracy Can Lead to Restrictive Immigration Policy. *Journal of European Public Policy, 19*(4), 1–19.

Howlett, M., Ramesh, M., & Perl, A. (2009). *Studying Public Policy: Policy Cycles and Policysubsystems.* Oxford: Oxford University Press.

Imbeau, L. M., Pétry, F., & Lamari, M. (2001). Left–Right Party Ideology and Government Policies: A Meta-Analysis. *European Journal of Political Research, 40*(1), 1–29.

Ireland, P. (2004). *Becoming Europe: Immigration, Integration and the Welfare State.* Pittsburgh: University of Pittsburgh.

John, P. (1998). *Analysing Public Policy.* London/New York: Continuum.

Kavanagh, D., Richards, D., Smith, M., & Geddes, A. (2006). *British Politics.* Oxford: Oxford University Press.

Keith, D., & McGowan, F. (2014). *Radical Left Parties and Immigration Issues* (SEI Working Paper No. 132). University of Sussex.

Kingdon, J. (1995). *Agendas, Alternatives, and Public Policies.* New York: HarperCollins.

Kirchheimer, O. (1966). The Transformation of Western European Party Systems. In J. LaPalombara & M. Weiner (Eds.), *Political Parties and Political Development* (pp. 177–200). Princeton: Princeton University Press.

Kriesi, H., Grande, E., Lachat, R., Dolezal, M., Bornschier, S., & Frey, T. (2008). Globalization and Its Impact on National Spaces of Competition. In H. Kriesi, E. Grande, R. Lachat, M. Dolerzal, S. Bornschier, & T. Frey (Eds.), *West European Politics in the Age of Globalization* (pp. 1–23). Cambridge: Cambridge University Press.

Lachat, R., & Kriesi, H. (2007). *The Impact of Globalization on National Party Configurations in Western Europe.* Paper Presented for the 1007 Annual Meeting of the Midwest Political Science Association, April 12–15, Chicago.

Lahav, G. (1997). Ideological and Party Constraints on Immigration Attitudes in Europe. *Journal of Common Market Studies, 35*(3), 377–406.

Lahav, G. (2004). Public Opinion Toward Immigration in the European Union: Does It Matter? *Comparative Political Studies, 37*(10), 1151–1183.

Lavanex, S. (2001). Migration and the EU's New Eastern Border: Between Realism and Liberalism. *Journal of European Public Policy, 8*(1), 24–42.

Levi, M. (1997). A Model, a Method, and a Map: Rational Choice in Comparative and Historical Analysis. In M. I. Lichbach & A. S. Zuckerman (Eds.), *Comparative Politics* (pp. 19–41). Cambridge: Cambridge University Press.

Lewis, M., & Grimes, A. (1999). Metatriangulation: Building Theory from Multiple Paradigms. *The Academy of Management Review, 24*(4), 672–690.

Lewis, M. W., & Kelemen, M. L. (2002). Multiparadigm Inquiry: Exploring Organizational Pluralism and Paradox. *Human Relations, 55*(2), 251–275.

Lowi, T. J. (1964). American Business, Public Policy, Case-Studies, and Political Theory. *World Politics, 16,* 677–715.

Lowndes, V. (2002). Institutionalism. In D. Marsh & G. Stoker (Eds.), *Theory and Methods in Political Science* (pp. 60–80). Basingstoke: Macmillan.

Mahoney, J., & Thelen, K. (2010). A Theory of Gradual Institutional Change. In K. Thelen & J. Mahoney (Eds.), *Explaining Institutional Change: Ambiguity, Agency, and Power* (pp. 1–38). Cambridge: Cambridge University Press.

March, J. G., & Olsen, J. P. (2006). Elaborating the "New Institutionalism". In R. A. W. Rhodes, S. A. Binder, & B. A. Rockman (Eds.), *The Oxford Handbook of Political Institutions* (pp. 3–23). Oxford: Oxford University Press.

McLean, I. (2001). *Rational Choice and British Politics*. Oxford: Oxford University Press.

Menz, G. (2008). *The Political Economy of Managed Migration: Nonstate Actors, Europeanization, and the Politics of Designing Migration Policies*. Oxford: Oxford University Press.

Meyers, E. (2000). Theories of International Immigration Policy – A Comparative Analysis. *International Migration Review, 34*(4), 1245–1282.

Money, J. (1999). *The Political Geography of Immigration Control*. New York: Cornell University.

Morales, L., Pilet, J. B., & Ruedin, D. (2015). The Gap Between Public Preferences and Policies on Immigration: A Comparative Examination of the Effect of Politicisation on Policy Congruence. *Journal of Ethnic and Migration Studies, 41*(9), 1495–1516.

Mudde, C. (2013). Three Decades of Populist Radical Right Parties in Western Europe: So What? *European Journal of Political Research, 52*(1), 1–19.

Odmalm, P. (2011). Political Parties and "the Immigration Issue": Issue Ownership in Swedish Parliamentary Elections 1991–2010. *West European Politics, 34*(5), 1070–1091.

Odmalm, P. (2014). *The Party Politics of Immigration and the EU*. Basingstoke/New york: Palgrave.

Odmalm, P., & Bale, T. (2015). Immigration into the Mainstream: Conflicting Ideological Streams, Strategic Reasoning and Party Competition. *Acta Politica, 50*(4), 365–378.

Perlmutter, T. (1996). Bringing Parties Back In: Comments on 'Modes of Immigration Politics in Liberal Democratic Societies'. *International Migration Review, 30*(1), 375–188.

Peters, B. G. (2012). *Institutional Theory in Political Science: The New Institutionalism*. New York: Continuum International Publishing.

Peters, B. G., Pierres, J., & King, D. S. (2005). The Politics of Path Dependency: Political Conflict in Historical Institutionalism. *The Journal of Politics, 67*(4), 1275–1300.

Pierson, P. (1993). When Effect Becomes Cause: Policy Feedback and Political Change. *World Politics, 45*(04), 595–628.

Pierson, P. (1996). The New Politics of the Welfare State. *World Politics, 48*(2), 143–179.

Pierson, P. (2000). Increasing Returns, Path Dependence, and the Study of Politics. *American Political Science Review, 94*(2), 251–267.
Pierson, P. (2004). *Politics in Time: History, Institutions, and Social Analysis*. Princeton: Princeton University Press.
Pierson, P., & Skocpol, T. (2002). Historical Institutionalism in Contemporary Political Science. In I. Katznelson & H. V. Milner (Eds.), *Political Science: The State of the Discipline* (pp. 693–721). New York: W. W. Norton.
Rhodes, R. A. W. (1986). *The National World of Local Government*. London: Allen & Unwin.
Rhodes, R. A. W. (1997). *Understanding Governance: Policy Networks, Governance, Reflexivity and Accountability*. Philadelphia: US Open University Press.
Richards, D., & Smith, M. (1997). How Departments Change: Windows of Opportunity and Critical Junctures in Three Departments. *Public Policy and Administration, 12*(2), 62–79.
Sabatier, P. A. (1988). An Advocacy Coalition Framework of Policy Change and the Role of Policy-Oriented Learning Therein. *Policy Sciences, 21*(2), 129–168.
Sanders, E. (2006). Historical Institutionalism. In R. A. Rhodes, S. A. Binder, & B. A. Rockman (Eds.), *The Oxford Handbook of Political Institutions* (pp. 39–56). Oxford: Oxford University Press.
Schain, M. (2008). Commentary: Why Political Parties Matter. *Journal of European Public Policy, 15*(3), 465–470.
Schmidt, M. (1996). When Parties Matter: A Review of the Possibilities and Limits of Partisan Influence on Public Policy. *European Journal of Political Research, 30*(2), 155–183.
Schmidt, V. (2006). Institutionalism. In C. Hay, M. Lister, & D. Marsh (Eds.), *The State: Theories and Issues* (pp. 98–118). Hampshire: Palgrave Macmillan.
Schwartz, H. (2001). *Down the Wrong Path: Path Dependence, Markets, and Increasing Returns* (Unpublished Manuscript). Charlottesville: University of Virginia.
Smith, M. (2006). Pluralism. In C. Hay, M. Lister, & D. Marsh (Eds.), *The State: Theories and Issues* (pp. 39–58). Hampshire: Palgrave Macmillan.
Smith, M. J. (1993). *Pressure, Power and Policy*. Hempel Hempstead: Harvester Wheatsheaf.
Smith, J. (2008). Towards Consensus? Centre-Right Parties and Immigration Policy in the UK and Ireland. *Journal of European Public Policy, 15*(3), 415–431.
Somerville, W. (2007). *Immigration Under New Labour*. Bristol: Policy Press.
Somerville, W., & Goodman, S. W. (2010). The Role of Networks in the Development of UK Migration Policy. *Political Studies, 58*(5), 951–970.
Statham, P., & Geddes, A. (2006). Elites and the "Organised Public": Who Drives British Immigration Politics and in Which Direction? *West European Politics, 29*(2), 248–269.

Statham, P., & Trenz, H. J. (2013). *The Politicization of Europe: Contesting the Constitution in the Mass Media*. London: Routledge.

Steinmo, S., Thelen, K., & Longstreth, F. (1992). *Structuring Politics: Historical Institutionalism in Comparative Analysis*. Cambridge: Cambridge University Press.

Streeck, W., & Thelen, K. (2005). Introduction: Institutional Change in Advanced Political Economies. In W. Streeck & K. Thelen (Eds.), *Beyond Continuity* (pp. 1–40). Oxford: Oxford University Press.

Surel, Y. (2000). The Role of Cognitive and Normative Frames in Policy-Making. *Journal of European Public Policy, 7*(4), 495–512.

Szczerbiak, A., & Taggart, P. (2008). *Opposing Europe? The Comparative Party Politics of Euroscepticism*. Oxford: Oxford University Press.

Thelen, K. (2004). *How Institutions Evolve*. Cambridge: Cambridge University Press.

Thelen, K., & Steinmo, S. (1992). Historical Institutionalism in Comparative Politics. In S. Steinmo, K. Thelen, & F. Longstreth (Eds.), *Structuring Politics: Historical Institutionalism in Comparative Analysis* (pp. 1–33). Cambridge: Cambridge University Press.

Tichenor, D. J. (2002). *Dividing Lines: The Politics of Immigration Control in America*. New Jersey: Princeton University Press.

Triadafilopoulos, T., & Zaslove, A. (2006). Influencing Migration Policy from Inside: Political Parties. In M. Giugni & F. Passy (Eds.), *Dialogues on Migration Policy* (pp. 171–193). Oxford: Lexington Books.

Van der Leun, J. (2006). Excluding Illegal Migrants in the Netherlands: Between National Policies and Local Implementation. *West European Politics, 29*(2), 310–326.

Vennesson, P. (2008). Case Studies and Process Tracing: Theories and Practices. In D. della Potra & M. Keating (Eds.), *Approaches and Methodologies in the Social Sciences: A Pluralist Perspective* (pp. 223–240). Cambridge: Cambridge University Press.

Watson, M., & Hay, C. (2003). The Discourse of Globalisation and the Logic of No Alternative: Rendering the Contingent Necessary in the Political Economy of New Labour. *Policy and Politics, 31*(3), 289–305.

Wilson, J. Q. (1980). *The Politics of Regulation*. New York: Basic Books.

Wright, C. (2010). *Policy Legacies and the Politics of Labour Immigration Selection and Control*. PhD dissertation, Cambridge University.

Wright, C. (2012). Policy Legacies, Visa Reform and the Resilience of Immigration Politics. *West European Politics, 35*(4), 726–755.

CHAPTER 3

Making the Migration State: The History of Britain's Immigration Policy

3.1 Introduction

Britain was once dubbed as a 'country of zero immigration' and not without cause (Freeman 1994, 297). Throughout the twentieth century successive governments, regardless of party affiliation, sought to limit colonial immigration on the assumption that good race relations necessitated minimum immigration. This was the bipartisan consensus that underpinned Britain's immigration policy for 50 years. Britain's highly restrictive approach has been widely acknowledged, but scholars have contested the reasons why. Given that the majority of Britain's post-war restrictive measures were targeted at non-white immigrants, many have rightly argued that Britain's immigration regime was underpinned by a racialised discourse (Paul 1997; Saggar 1992; Spencer 2002; Hampshire 2005). Other scholars, namely Randall Hansen (2000), challenged the racialisation thesis, claiming that the political elite was rather responding to electoral demands for less immigration (Hansen 2000, 263). Notwithstanding the reasons why policy evolved as it did, Britain's immigration regime was undoubtedly a restrictive one.

The Labour governments of 1997–2010 pursued an expansionary economic immigration policy in stark contrast to Britain's past record. The policy shift under the Labour governments was akin to a third order change (Hall 1993), where the fundamental objectives and framing of immigration policy fundamentally deviated from what had hitherto been the case. Labour's immigration policy was a historically unprecedented

© The Author(s) 2018
E. Consterdine, *Labour's Immigration Policy*,
https://doi.org/10.1007/978-3-319-64692-3_3

reversal and this book seeks to explain how and why this happened. But in order to demonstrate the degree of policy change, we need contextual background. This chapter outlines Britain's post-war immigration policy and polarises this legacy with a comprehensive overview of Labour's immigration policy. The chapter builds a narrative of British immigration policy from 1948 until 2010 to demonstrate the unprecedented shift under the Labour governments in comparison to Britain's post-war restrictive framing. The chapter begins by summarising the political debates and the main policies and acts that were established in the 1960s, 70s, 80s and 90s, demonstrating both the degree of restrictionism and the lack of any strategic immigration policy. The chapter goes on to provide a detailed narrative of the reforms made under the Labour administrations, with a particular focus on Labour's expansive economic immigration programme.

3.2 Maintaining Fortress Britain: 1948–1970

In contrast to other Western states that facilitated immigration through formal programmes to varying degrees (Layton-Henry 1992, 12), post-war Commonwealth immigration to Britain was largely spontaneous and always unwanted. For the first 'three decades of the post-war period, polling data reveal consistent majority public opposition to New Commonwealth migration' (Hansen 2000, 4), archival sources show uneasiness in Whitehall over non-white migration, and the scholarly literature, without exception, stresses the hostility and racism of successive British governments towards New Commonwealth migration (Spencer 2002; Hampshire 2005; Messina 1989; Layton-Henry 1992; Paul 1997).

Although migration policy had been informed by two other Acts pre 1945—the Aliens Act of 1905 and the Aliens Restriction Act of 1914—it was only post 1945 that 'immigration beyond Europe became significant enough to register as a major political issue' (Hampshire 2005, 9). Previously, citizenship had been derived from a common code of British subjecthood, but as Commonwealth countries began to gain independence, it was evident that this overarching mode of citizenship was no longer sustainable. The government had to negotiate the process of decolonialisation whilst maintaining the doctrine of equal rights for all British subjects, a principle at the very core of Commonwealth identity. Thus Britain's immigration policy essentially began in 1948, with the establishment of the British Nationality Act (BNA), which conferred British subject status—Citizen of the United Kingdom and Colonies (CUKC)—to all members of the Empire (an estimated 600 million people), serving as a last attempt to reaffirm Britain as the leader of the Commonwealth, 'to reinforce a notion of imperial unity wobbling under the impact of decolonization' (Carter et al. 1987, 142). As the Minister of State for Colonial Affairs in the Churchill Government proudly told the House of Commons in 1954: 'In a world in which restrictions on personal movement and immigration have increased we still take pride in the fact that a man can *say civis Britannicus sum* whatever his colour may be, and we take pride in the fact that he wants and can come to the mother country' (Parliamentary Debates (commons) (532), col. 827, 5 November 1954).

Yet the BNA had a constitutional purpose and was not expected to facilitate immigration. Extraordinarily, at no stage in the debates over the Bill was the possibility entertained that substantial numbers of colonial citizens could, or would, exercise their right to permanently reside in Britain, although prominent Conservatives in the House of Lords did suggest that attempts to define a Commonwealth citizen would lead to division and legal arguments (Dean 1987, 317).

Following Britain's post-war reconstruction, initially this all-encompassing citizenship proved to be advantageous for Britain's labour market. However, Britain's aspiration to preserve some hold on the empire came with some unexpected and unwanted consequences. Following the BNA, a wave of unanticipated Commonwealth migrants arrived on Britain's shores, symbolised by the infamous arrival of the Empire Windrush in 1948, although it should be noted how comparatively small such flows were in contrast to European migrants (Carter et al. 1987, 147). Nonetheless, Black and Asian immigrants' permanent settlement in Britain was undoubtedly viewed as a problem by the political elite, supported by the Royal Commission on Population (Layton-Henry 1992, 28) and the independent Political and Economic Planning institute who claimed that 'the absorption of large numbers of non-white immigrants would be extremely difficult' (Political and Economic Planning, "Population Policy in Great Britain", 1948). Yet immigration legislation, the elite argued, would undermine Commonwealth unity. As a result, during the 1950s the government attempted to limit colonial immigration through administrative measures, including incentives to colonial governments to limit the issuance of passports and travel documents, along with wider dissuasion tactics to discourage colonial immigrants from coming to Britain (Spencer 2002, 31). As Ian Spencer speculates, this was 'defacto immigration policy' (Spencer 2002, 8).

Whilst Black and Asian settlement remained an unwanted presence, Britain's international standing in the Commonwealth deterred the government from pursuing legislative action. In short, neither the public or political elites deemed that the 'problem' was serious enough yet to pay the political price for restrictions on colonial immigrants (Carter et al. 1987, 147). Alongside such developments, as migration was concerned the doctrine of *Civis Britannicus sum* symbolised a commitment to movement between the Old (white) commonwealth and Britain. Thus as Old Commonwealth countries pursued their interests within a regional rather than Commonwealth framework, so support for the doctrine declined, and in turn the barriers to immigration controls were loosened.

As rumours spread that immigration controls were imminent, a record net immigration of 191,100 was recorded for 1960–1961, more than for the previous five years combined (Layton-Henry 1992, 13). In turn, the Cabinet committee on colonial immigrants warned in a memo that 'the movement was reaching a stage at which the government would be obliged to introduce legislation to enable them to control it' (TNA CAB 128/35, CC (61) 29, 30 May 1961). By 1961 the government conceded that this was an untenable situation in need of legislation, which came in the form of the 1962 Commonwealth Immigration Act (CIA).

The 1962 Act severely curtailed primary immigration by establishing a labour-voucher system, marking a watershed moment for Britain and the Commonwealth as a whole—for the first time the right of British subjects to enter the 'mother country' was restricted. Exemption from immigration controls was made on two grounds: place of birth or the issuing authority of a CUKC passport (holders of a CUKC passport issued under the authority of London were exempt). Whilst the government faced real political difficulties at this stage, 'it is hard to disagree with the claim that racial attitudes underlay the differential treatment' (Hampshire 2005, 29), in particular that the Irish were exempt from these controls reflects the racist underwriting behind this legislation.

While primary immigration decreased as a result of the 1962 Act, secondary immigration did not (Layton-Henry 1992, 12), and by 1967 settlement of New Commonwealth citizens had greatly increased. Propelled by fears of so-called overcrowding (TNA, HO 344/42, 9th November 1961; TNA, HO 344/42, 9th November 1961), this surge caused concern amongst the political elite (TNA HO 344/193, undated.). Meanwhile, events outside of Britain left them in an even more contentious position.

Turbulent times were occurring in Africa; the Kenyatta government had begun an aggressive Africanisation policy, where Kenyan residents without African descent were persecuted and expelled from the country. As a result, Kenyan Asians who possessed CUKC status who were exempt from 1962 CIA controls because of their passports being issued by a London authority fled to Britain in fear of being left stateless. Troubled by an already antagonistic public, this unexpected wave of immigration caused panic amongst the political elite. In turn, Home Secretary James Callaghan sought approval from Cabinet to introduce further legislation, including the withdrawal of the right of Kenyan CUKCs to enter Britain. He argued that it was 'both urgent and essential' to extend controls to those 'who did not belong to this country in the sense of having any direct family connection with it or having been naturalized or adopted here' (TNA, CAB 128/43, CC 68 (13), 15 February 1968). The pressures on social services, Callaghan argued, 'would be such that large additional expenditure would be required, and our race relations policy would be in jeopardy' (TNA, CAB 128/43, CC (68) 14, 22 February 1968). 'Wide support', he assured ministers, 'could be expected in this country for a policy on these lines' (TNA, CAB 128/43, CC (68) 14, 22 February 1968). Subsequently, and given that the government left up to 200,000 people effectively stateless, most would say shamelessly, the government passed the 1968 Commonwealth Immigrants Act (CIA), an Act that was 'loathed by liberal opinion and loved by the public' (Hansen 1999, 821).

It is worth highlighting that towards the end of the 1960s, there was an attempt to frame immigration policy on economic grounds. Purportedly 'there [was] an established general move away from the emotive atmosphere that has surrounded this question' (TNA Lab 8/3315, 16 May 1966), when research commissioned by the Home Office revealed that immigrants were less costly to the public purse than the indigenous population (TNA HO 376/130, undated; TNA HO 344/63, undated). However, the suggestion that immigration could be relegated on the basis of economic needs was refuted in 1970, denigrated as an irresponsible approach to policy and it was resolved that 'justification by GNP is a pitfall ridden approach' (TNA HO 376/130, 21 December 1970).

3.3 1970–1980: Joining the Community

The rising political saliency of immigration in the late 1960s meant that there was potential for the 'race card' to be played in the 1970 General Election (Saggar 2000). For the first time, immigration was the fourth most salient issue for voters in the Election (Geddes 2008) This was in no small part due to Conservative MP Enoch Powell's contribution to the debate, when he infamously made his 'rivers of blood' speech in 1968. While Powell was sacked immediately from the shadow cabinet, 'there is little doubt that Heath accepted that the public support enjoyed by Powell necessitated a greater restrictionism in Conservative policy' (Hansen 2000, 190). In turn, the Conservative Party toughened up their rhetoric and pledged in their manifesto to give the Home Secretary 'complete control over the entry of individuals into Britain', promising that 'there will be no further large scale permanent immigration' (Craig 1990, 127). The public rewarded the Conservatives at the 1970 General Election (Layton-Henry politics p. 84) with an unexpected victory, gaining an estimated increment of 6.7 per cent in votes because many 'perceived them to be the party more likely to keep immigrants out' (Studlar 1978, 46).

With an electoral pledge to curtail further immigration, action was needed but there was at the same time an imperative interest in preserving some hold of the Commonwealth, as Home Secretary Reginald Maudling argued, '[there is a case] on kith and kin grounds, for special provision for those Commonwealth citizens who have an ancestral connection with the UK' (TNA CAB 129/154, cp (70) 126, 31 December 1970). Yet it was difficult to achieve this without 'discriminating between different members of the Commonwealth' (TNA CAB 129/154, cp (70) 126, 31 December 1970). The panacea was to pass an Act that included an exemption through patriality for those with a grandparent or parent born or naturalised in Britain, a mechanism 'clearly designed to secure access for Australians, Canadians and New Zealanders while denying it to the rest of the Commonwealth' (Hansen 2000, 195). The status quo was retained; the objective to limit unassimilable Colonial immigrants persisted.

The 1971 Act represents the final deterioration of universal Commonwealth citizenship and essentially 'went as far as it could in explicitly diminishing the former privileges of Commonwealth citizens' (Cohen 1997, 365). Furthermore, the Act gave new unfettered powers to the Home Secretary by bestowing legal authority to grant leave to remain and make immigration rules. Such power remains in place today. The Act also gave new powers to deport Commonwealth migrants and created new barriers so that Commonwealth migrants would need a work permit for a specific job, with only skilled immigrants being issued a permit. In effect, this meant that the criteria for permits for Commonwealth citizens would be stiffer than for aliens (Hampshire 2005, 39):

> Through this reconstruction of subjecthood, the Act legally differentiated between the familial community of Britishness composed of the truly British-those descended from white colonizers and the political community of Britishness composed of individuals who had become British through conquest or domination. The latter community discovered that as a result of the 1971 Immigration Act, their British nationality amounted to little more than a name on a passport and that their access to Britain was restricted in much the same way as it was for aliens. (Paul 1997, 181)

As controversial as the Act was, as a control mechanism it was redundant; the previous two statutes had effectively halted New Commonwealth flows (Spencer 2002, 143), and settlement had already decreased by roughly 1,000 between 1969 and 1972 (TNA: HO 344/63, undated.). In turn, the Act was criticised for being a 'sop to racial prejudice' because it would have little effect on numbers (Parliamentary Debates (Commons) (813), cols. 120–4, 8 March 1971) and faced heavy criticism from the Labour Party (TNA: HO 344/97, 1972)—somewhat hypocritically given that 'the blatantly racist aspects of the patriality clauses' of the Act were 'foreshadowed' in Labour's own 1968 CIA (Layton-Henry 1992, 89)— and NGOs (TNA: FCO 50/484, 14 December 1972) both on grounds of racial discrimination and the new unfettered powers of the Home Secretary:

> It charts the dangers of injustice and abuse to be found in an executive discretion that can, and does, operate outside, and even in contradiction to, the recommendations of the Courts. It is a discretion almost totally unfettered by independent scrutiny and which lies beyond the safeguards of liberty. (Prisons Reform Trust, *A Law Unto Themselves: Home Office Powers of Detention* (London: The Trust, 1984, 1)

Aside from criticism within Parliament, the Act also generated interdepartmental tensions, with the Treasury being particularly opposed, being concerned that the additional manpower needed as a consequence of the Act, would be disproportionate to the outcomes:

Political factors apart, the papers do not seem to me to advance any arguments which would justify incurring additional public expenditure of the order of £400,000 or £500,000 a year at a time of financial stringency. However, given that the decision has been made on political grounds, I do not think there is any point of detail to which we should object. (TNA: T227/3216, 4 December 1970)

The DEP similarly raised concerns stating that they could not meet the extra cost within their public expenditure budget. At the heart of the Treasury's objection was that the Bill would not actually achieve its objective of reducing Commonwealth immigration and that flows were likely to remain the same. Nonetheless, the 'feeling in the [Conservative] Party [was] for no more immigration at all…it has proved impossible to reassure the public' (TNA: T 353/68, 6 March 1973). Thus despite widespread opposition, the Act passed and remains the principal statutory instrument to regulate immigration today.

Notwithstanding the general concern to limit New Commonwealth immigration, the 1971 Act was partly motivated by the 'numbers game'. With 1973 came a pivotal and defining moment for Britain, acceptance to the European Economic Community (EEC). While the Home Office repeatedly claimed that immigration flows from the EEC would be small (TNA CAB 164/460, 16 February 1970), it was found that prior to accession Italian migrants had obtained a large number of work permits. In light of this impending wave of now permit-free labour, Heath rejected a proposal to increase the issuance of work permits from 3,500 to 5,500 on the grounds that this figure was unjustifiably high (TNA FCO 50/484, 23 January 1973). Heath suggested that the dependence on foreign workers in the catering and hospitality trade in particular, should be reduced. The Treasury were critical of Act claiming that the additional staff needed to enforce the Act would incur a public expenditure of £500,000 a year (TNA T227/3216, 4 December 1970). The Treasury nonetheless accepted that the Act had a political purpose. The Treasury did however demand an exemption of overseas doctors (TNA T227/3216, 17 August 1970), to which the Home Office conceded.

As accession approached in the early 1970s, now familiar fears of EEC nationals coming to Britain to 'welfare shop' were also on the horizon (TNA CAB 164/460, undated). First and foremost being both the leader of the Commonwealth and an EEC Member State now put the government in a geopolitically awkward position as 'It would be politically untenable to put Commonwealth citizens in general so much at a disadvantage with foreigners, but socially disastrous to throw the doors as widely open to Commonwealth citizens in general as to nationals of other member states' (TNA CAB 164/460, undated). This proved politically problematic for the government; asked whether priority would be given to EEC nationals over Commonwealth citizens, 'consistently ministers' replies sought to duck the issue' (TNA FCO 50/358, undated). With a potentially large pool of unlimited EEC immigration imminent, it was resolved that Commonwealth migration, or more specifically New Commonwealth migration, must be limited further. The Act was a symbolic attempt to redefine British citizenship at a time where its imperialistic power in the Commonwealth was diminishing. Thus the 1971 Immigration Act was, in part, a manifestation of Britain's conflicting geopolitical interests.

Underlying the restrictive Acts throughout the post-war period was a bipartisan consensus that good race relations required strict immigration controls. Fractious race relations had been a concern amongst the political elite since the mid-1960s when race riots broke out across the country, and there was consequently an acceptance of the need to challenge widespread racism. This dual approach of restrictionism whilst challenging discrimination is what Shamit Saggar (1992) coined the 'Hattersley equation', named after Labour Home Secretary Roy Hattersley (Rex and Tomlinson 1979). The 1965 White Paper *Immigration from the Commonwealth* (HMSO 1965) was the first official articulation of this dual approach, but it was the establishment of anti-discrimination laws in the form of the 1965 Race Relations Act, the 1968 Race Relations Act and the 1976 Race Relations Act, which institutionalised the 'Hattersley equation'.

Much like the 1960s, there was little in the way of any type of economic immigration policy. Whilst it was suggested in 1976 that it would be 'useful to see what an optimum immigration policy might be if the United Kingdom could start with a clean sheet', it was ultimately conceded that 'that there was no possibility of preparing long-term forecasts of the labour market as a basis for immigration policy' (TNA CAB/130/882, 23 September 1976). By 1976 it was becoming apparent to the government that there was no coherent immigration policy as such: 'Present policy represents a series of accretions, the different elements of policy being the response to pressures from immigration from different areas over recent years...there is no coherent criteria for policy' (TNA CAB 130/882, 20 December 1976).

The running theme of the 1970s was restrictionism by any means, and the Labour government largely believed that they had succeeded in ending all large-scale primary immigration from New Commonwealth countries and Pakistan (TNA CAB 130/1007, 1978). Indeed it was claimed that the argument about numbers of immigrants 'was over' (TNA CAB 130/1007, 1978). But whilst the government appeared confident that primary immigration from New Commonwealth countries had ended, public concerns over immigration were still rife and the government feared that 'almost anything said to the press on this delicate subject at present was liable to cause misunderstanding and, in effect, to continue to maintain the temperature of a subject which it was in all our interests to cool' (TNA FCO 50/585, 11 June 1976).

It was deemed necessary for further restrictions to be put in place, especially as further EC enlargement (Southern Europe) could 'affect the number of work permits issued to Commonwealth immigrants' (TNA FCO 30/3967, 22 December 1977), and there were subsequently numerous discussions as to how to achieve this. The introduction of a points-based system was even discussed but ultimately whilst the model had 'its attractions, in the present economic circumstance the additional burdens upon manpower and financial resources would probably be unacceptable' (TNA FCO 37/1832, 2 February 1977).

3.4 1980–1997: Swamping Britain and the Asylum Crisis

The 1980s and 1990s were relatively dormant in terms of immigration legislation or policy. Nonetheless, the period maintained the status quo of restriction. Prior to the 1979 Conservative landslide, Thatcher (1978) made her infamous 'swamping' statement about the impact of immigration on British culture. This statement signified that race would no longer be regarded by the Conservative Party as an untouchable issue (Messina 1985, 425). The Conservative governments continued restrictionism through the 1981 British Nationality Act, which introduced a definition of citizenship exclusive to British citizens, and was designed 'as with earlier rules, to control immigration for work and settlement' (Hansen 2000, 207; TNA FCO 107/30, 1979). The Act also included the Primary Purpose Rule that severely restricted family reunification rights. Essentially, 'insofar as Britain can be said to have an immigration policy, it [was] a policy designed to contain the social problems of past immigration by eliminating virtually all future inwards flows' (Rees 1982, 95).

The 1980s was much like the 1970s, inactive in terms of any type of economic immigration policy with one exception; the work permit scheme underwent two reviews. By 1980, work permits for full time employment were only available to overseas workers holding recognised professional qualifications or having a high degree of skill and/or experience. The system no longer accommodated large influxes of semi-skilled or unskilled workers. The first review in 1981 aimed to improve the efficiency of the work permit system, but only minor amendments were made (Salt and Kitchling 1990, 270).

In contrast to 1981, the 1989 review took place under considerably better economic conditions. The review shifted the policy focus 'from employment protection towards encouragement of enterprise economy' (Clarke and Salt 2003, 564). This led to changes in the work permit system in 1991, where a two-tier system for processing applicants was established. While these reviews led to fairly fundamental changes to the work permit system, the changes made were to simplify entry, rather than encourage or liberalise immigration and there was 'no relaxation of the conditions [to obtain a work permit]' (Hansard 31 January 1991, col632).

With the 1990s came a new dilemma for the political elite, the 'asylum crisis'. As a result of political changes worldwide, people across many nations were fleeing their countries in fear of persecution, and subsequently asylum applications to Britain greatly increased, indeed an 'unprecedented increase' according to Home Secretary Kenneth Baker (Hansard 7 February 1991, col 400). As the political saliency of asylum seeking increased, so did the government's efforts to control humanitarian immigration in the form of primary legislation. Four parliamentary acts on asylum and immigration were established under the Conservative governments in the 1990s, including the 1993 Asylum and Immigration Appeals Act, which was the first asylum-specific legislation in Britain. The Act introduced compulsory fingerprinting of asylum seekers and extended the 1987 Carrier's Liability Act. While the 1993 Act was effective in limiting granted asylum appeals by a dramatic amount—six months prior to the Act 86 per cent of asylum applicants were granted either asylum or exceptional leave to remain, compared with 28 per cent in the six months following the Act—asylum applications continued to rise. The Conservative government then implemented the 1996 Asylum and Immigration Act, which removed certain in-country appeal rights and withdrawal of social security benefits for asylum seekers (Stevens 1998, 207). Yet these provisions did little to alleviate the now colossal backlog of asylum applications in Britain. Enter a revitalised Labour government to take a stab at resolving this now highly charged political situation.

A brief look at 50 years of policy developments clearly demonstrates how and why Britain got its reputation for being a 'country of zero immigration' (Freeman 1994). The bipartisan consensus underpinning policy was the belief that good race relations necessitated limited immigration. Unchallenged and widely accepted by all parties, this logic was embedded in every policy throughout the twentieth century. Britain began with an unintentionally open door policy and subsequently spent the next 50 years trying to close it and ultimately end Britain's 'debt to history' (TNA CAB 130/882, 17 September 1976). It presents a picture of reactive policymaking, in an effort to defuse the saliency of the issue and above all else curtail almost all types of immigration.

Whilst issues of citizenship, race relations and asylum legislation all received attention from the British political elite, there was no real economic immigration policy to speak of. Of course the absence of a policy could be regarded as a policy in itself, if only by adopting a defusing strategy (Bale et al. 2010). Migrants had filled labour shortages throughout the post-war period, and there were a handful of meagre attempts to explicitly recruit foreign labour following the War. But the decision governing such entries was 'quintessentially a political one' (Hansen 2000, 10). Economic or labour immigration was not regarded as an autonomous policy, and no politician would dare suggest that such policy could or should be regulated by the needs of the economy, thus 'it is fair to say that there had been little attempt to use immigration or labour migration in a systematic way to counter labour or skills shortages' (Caviedes 2010, 110). Yet Britain's new, revitalised, "cool Britannia" government was about to change all of this.

3.5 Immigration Policy Under New Labour

'Managed migration is not just good for this country. It is essential for our continued prosperity' (Tony Blair 2005)

Major changes were apparent in all areas of immigration policy under the Labour administrations, but arguably economic immigration received the most significant transformation, marking a '…decisive break with the previous policy model'; see appendix 1 for an overview (Somerville 2007, 29; Flynn 2005). From 1997–2010 Britain's immigration system underwent substantial changes including: ten major parliamentary acts on immigration and asylum, countless strategy documents, and major reforms and renovation of the immigration system (Somerville 2007, 24). Whilst the immigration regime in this period remained restrictive to some types of immigration (such as asylum and irregular migration), the Labour governments' economic immigration reforms culminated in one of the most expansionary policies in Western Europe.

During their time in office, the Labour administrations constructed a system of selective admission based on the economic utilisation of migrants. With the exception of temporary labour voucher schemes to fill labour market shortages in the 1960s—which was predominantly established in a bid to control and regulate unwanted Commonwealth immigration—this was the first attempt at a strategic economic immigration policy. These policy changes had a cumulative impact on immigration flows, with net migration to the UK reaching an annual average of 195,000 during 2000–2011, which was three times the annual average of 65,000 in the previous decade (Vargas-Silva 2013). With two and a half million foreign-born workers added to the British population during Labour's time in office, it is no exaggeration to say that immigration, under the New Labour administrations, has 'permanently changed the face of Britain' (ONS 2010; Finch and Goodhart 2010, 4).

3.6 Dealing with the Asylum Crisis: 1997–2000

While the Labour Party radically reformed economic immigration policy over their time in government, there was no abrupt shift upon electoral victory, nor did the Party enter office with a thought-out strategy for immigration reform (Layton-Henry 2004). On the contrary, immigration did not feature in Labour's election campaign, and the only reference to it in the 1997 manifesto was a commitment to abolish the primary purpose rule governing marriage migration, on the grounds that it was 'arbitrary' and 'unfair' (Labour Party 1997), a pledge that was met within a month of taking office. Immigration was simply not on the political agenda.

When the Labour Party entered government in 1997, it was faced with three pressing immigration issues. First, a strong public perception that administrative procedure (predominantly incompetence in the Immigration and Nationality Directorate) was in a chaotic state. Second, the asylum backlog had 'become the immigration scandal of the late 1990s in its own right' (Clayton 2008, 14). Third, the government was faced with the decision as to whether to sign the Amsterdam Treaty (Flynn 2005, 471).

The Treaty included provisions to allow legal competency in the field of immigration and asylum under the first pillar. Due to concerns over the Schengen acquis, the UK negotiated a special opt-out protocol, allowing flexibility as to which asylum and immigration proposals they would opt in/out of. Since its acceptance into the community, Britain was a sceptical partner and thus reluctant to move towards a common EU immigration and asylum policy, opting out of most directives. Amsterdam was no exception, as Foreign Secretary Robin Cook declared, 'Policy on border controls and immigration will be made in Britain, not in Brussels' (Hansard 12 November 1997, column 910).

While the UK has adopted measures in the realm of race equality, workplace discrimination, asylum acquis and border controls, Britain has consistently opposed common European policies that would empower supranational institutions, which 'could challenge the executive's tight grip on immigration policy' (Geddes 2003, 31). Britain's approach to EU regulations on immigration has 'enabled it to maintain a strongly national approach when it chooses to, while opting to collaborate where that helps to achieve its objectives, as on asylum and irregular migration' (Spencer 2011, 17). There is very limited evidence to suggest that the EU was a source of inspiration for Labour's approach to economic immigration. The 1999 Tampere Summit provided a five-year plan for immigration and asylum that included an objective of 'management of migration flows', yet the presidency conclusions made no mention of opening up economic immigration routes (Balch 2010, 128). It was not until 2000 (by which point the expansion of economic immigration had begun in the UK) that the European Commission began to talk of a new approach to labour migration, but arguably there was little substance or appetite to such proposals in any case (CEC 2000, 3).

The Labour government's first White Paper on immigration, *Fairer, Faster, and Firmer: A Modern Approach to Immigration and Asylum*, was published in 1998 (Home Office 1998). The paper set out the modernised efficiency approach that the government sought to apply to immigration and asylum policy, and there was very little in the way of strategic thinking on economic immigration (Flynn 2003, 4). Indeed, there were visible continuities between this paper and the bipartisan consensus of the linkages between control and race relations (Balch 2010, 124). The 1999

Immigration and Asylum Act swiftly followed, which included the paper's outline of dispersing asylum seekers to areas where accommodation costs would be low.

It is fair to say that Labour's first term was overshadowed by the asylum crisis. With an eightfold increase of asylum applications between 1988 and 1998, and a consequential backlog of 52,000 asylum cases from the previous administration (Hansard 27 October 1997, col 569–71; Hansard 27 July 1998 col 37–54), dealing with asylum was clearly the overriding concern for government. The somewhat controversial measures placed to combat the backlog, such as the 1999 Act, appeared to be successful in policy terms, as by the end of their second term asylum applications had greatly reduced. Border controls, asylum backlogs and racial equality all received considerable attention; economic immigration policy however was certainly not on the political agenda (at least publicly) at this stage.

3.7 2000–2005: THE MAKING OF MANAGED MIGRATION

Labour's second term of office saw a new Home Secretary—David Blunkett—who was the most assiduous minister in terms of developing an economic immigration strategy. The government continued to manage the asylum crisis but began focusing their attention on tackling irregular migration. In 2001 the total unauthorised migrant population was estimated to be at 430,000, causing alarm bells in Whitehall (Woodbridge 2005). Events such as the Sangatte Crisis in July 2001—when it was leaked that a group of Sangatte migrants managed to stage a break-in at Coquelles and travel several miles down the Channel before being caught—only heightened the salience of asylum and irregular migration. The Sangatte Crisis and 9/11 (and the subsequent anti-terrorism measures), combined with a pan-European concern over irregular immigration, resulted in a new negative framing of immigrants—that of a security risk (Huysmans 2000).

Yet behind the headlines and the multiple Commons debates on border control, asylum and security, the Labour government was establishing an economic immigration policy. Within this period, a distinction was built in rhetoric and policy between asylum on the one hand and economic immigration on the other. The separation between 'wanted' and 'unwanted' immigration allowed the Labour government to be tough on asylum whilst almost tacitly pursuing expansive economic immigration (Mulvey 2011, 1478).

Where hitherto the immigration system was restrictive to all types of immigration, it is within Labour's second term of office where the government essentially re-framed the immigration debate:

> In the mid-1990s there was a restrictive rhetoric, with the main aim to keep those coming to work and live in the UK to "an irreducible minimum". By the early 2000s this had been replaced by a more expansive logic of labour migration being used to "boost the economy". (Balch 2009, 625)

While it is difficult to pinpoint the defining shifting moment in strategy and policy, late 2000 to early 2001 was when concrete policy changes were made.

3.7.1 Our Competitive Future: High-skilled Migration

The beginning of this new strategic approach to immigration policy was first hinted at in the Department for Trade and Industry's 1998 White Paper *Our Competitive Future: Building the knowledge driven economy* (DTI 1998). Two years later, a major government review on immigration and its impact on the economy was conducted, which was strongly influenced by Minister for Competitiveness Alan Johnson. The impetus for this review was a wider re-thinking of global economic competitiveness, an initiative driven by the Treasury and the DTI. As a consequence of this review, two new labour immigration routes were established that were 'major departures from previous economic migration policy'—the Innovators Scheme (in 2000) and the highly skilled migrant programme (HSMP) in 2001 (Somerville 2007, 33). For the first time in Britain's history, the immigration system was geared towards the supply of skills as opposed to only demand.

Announced in July 2000 by Minister for Immigration Barbara Roche and Minister for Competitiveness Alan Johnson, the Innovators Scheme was a small-scale pilot targeted at attracting entrepreneurs to the UK. The justification for the scheme, given by Johnson, was that:

> In the knowledge-driven economy, innovation is ever more critical to success. We are committed to ensuring that by 2002—the UK will be the best place in the world to conduct e-commerce. This scheme will strengthen our

position in the global war for talent. It will promote the UK as the location of choice for high-tech entrepreneurs. (DTI 2000)

The two-year pilot aimed to pull in 2,000 applicants for each year of the project. However, the scheme attracted only 112 applicants, largely because of the creation of the HSMP. The HSMP (originally called the Skilled Migrant Entry Programme) was introduced in January 2002 and was initially planned as a 12-month programme. The scheme was 'designed to allow individuals with exceptional personal skills and experience to come to the UK to seek and take work' (Home Office 2002b). The HSMP was the first origins of the points-based system (PBS), with candidates being admitted according to the number of points for human capital (such as qualifications, work experience, past earnings). The initial pass rate was set at 75 points, but over the years of its operation the threshold for entry was eased; the pass mark was reduced to 65 points in 2002, the number of points prospective applicants could receive for previous work experience was doubled in January 2003, and new points categories such as being under 28 years old were added in October 2003 (Home Office 2002b; Somerville 2007, 33). By 2007, before the replacement of the scheme with Tier 1 of the points-based system, annual entry on the HSMP had reached 28,000 (Salt 2009).

As part of the launch of the schemes, Immigration Minister Barbara Roche made a landmark speech to the Institute for Public Policy Research (IPPR) in 2000, suggesting a fundamental change to the policy objectives of immigration. This was the first time that a minister had ever publicly expressed that immigration should be seen as part of the economic growth strategy:

> The evidence shows that economically driven migration can bring substantial overall benefits both for growth and the economy…Migration policy needs to be joined up—we need to recognise its importance for the economy, skills, employment, trade, investment, international relations, higher education and culture. (Roche 2000)

Within a year of this speech, and as part of the review process, the first major cross-government research paper was published in 2001. Conducted by economists and experts in the field, the now infamous report (see Chap. 6 for discussion) concluded that 'overall migration has the potential to deliver significant economic benefits' (Glover et al. 2001, (i)). A further report by the Home Office, Performance and Innovation Unit (PIU) and IPPR looking at the fiscal effects of migration concluded further positive gains from immigration (Gott and Johnson 2002).

3.7.2 Work Permits

Perhaps the most significant move towards liberalisation by the Labour government was the expansion of work permits. The demand for work permits had been rising throughout the 1990s, with a 10,000 increase in permits issued between 1987 and 1992 (Salt 2001), but it was the Labour governments that significantly expanded this scheme. According to the government the liberalisation of the scheme was made 'In response to growing concerns about skills shortages' (Home Office and DTI 2002, 7); thus the government refocused 'the work permits criteria to facilitate the easier inward migration of those with key skills in relation to the UK economy' (Home Office and DTI 2002, 7). The changes to the work permit scheme were in response to the 1999 Budget that 'argued for a loosening of the rules limiting the skills and experiences required for inward migration, especially for entrepreneurs and investors wishing to start businesses in the UK' (Home Office and DTI 2002, 26). Labour's liberalisation of economic immigration is clearly expressed in the numbers arriving on the work permit scheme, with numbers rising from approximately 24,000 in 1995 to a peak of 96,740 in 2006 (Salt 2009, 89).

The government further lowered the criteria to obtain a work permit in 2000, from a qualification and two years' work experience, to just the qualification. Moreover, applicants without a degree had previously needed five years work experience to obtain a work permit. This was lowered to three years in October 2000. The validity of a work permit was also extended from four to five years, and exceptions to the Resident Labour Market Test (RLMT) were introduced, such as for board-level managers. This was coupled by an easing of the rigidity of the administration of work permits, including intra-company transfers and multiple-entry work permits, the process of which had begun in the late 1990s (Ibid; Sommerville 2007, 30).

Employers had become disgruntled with the cumbersome and expensive bureaucratic processes to obtain work permits in the late 1990s, and in response the Department for Education and Employment (DfEE) initially took steps to make more rapid decisions on applications and to reduce the skill threshold for posts eligible for a permit. Part of this easing of administrative burdens was to transfer work permit-related passport endorsement practices from the infamously chaotic Immigration Nationality Directorate (IND) to Work Permits UK to prevent delays. This transfer proved to be an effective remedy, with Work Permits UK reputedly completing 90 per cent of applications within 24 hours by 2003 (Spencer 2011, 89). Moreover, the rules were relaxed for senior-level ICTs, board-level posts and those associated with inward investment (Home Office and DTI 2002, 26). These actions saw the number of permits (swelled by recruitment of IT and health professionals) rise to more than 85,000 in 2000 (Spencer 2011, 85). The cumulative expansion of the work permit scheme over these years resulted in a 41.8 per cent increase in applications by the end of 2000, with the number of work permit holders and their dependants reaching a record high of 137,035 in 2005 (Dobson et al. 2001). Between 1995 and 2002 total applicants for work permits (including HSMP applicants) had increased by an astonishing 300 per cent (Clarke and Salt 2003, 565).

3.7.3 Low and Semi-skilled: SAWS, SBS and Working Holidaymakers

At the other end of the labour market, the government expanded low-skilled immigration routes, such as the expansion of the Seasonal Agricultural Workers Scheme (SAWS). The long-running SAWS was set up after the Second World War to facilitate the movement of young people from across Europe to work in agriculture. Quotas had long been used to manage the number of people able to participate in the scheme, and throughout the 1990s this quota was set at 10,000 places. Under the Labour governments, however, the SAWS quota was increased by 15,000 places in total, rising to 15,200 in 2001 and 25,000 in 2003, because of 'shortages in the supply of seasonal and casual labour' (Work Permits UK 2002).

In other sectors of the UK economy, labour market shortages were becoming apparent. The hospitality sector had been persistently experiencing labour market shortages in the late 1990s (Burns 2003), and hotels and restaurants were consistently experiencing over 3 per cent vacancies (ten out of twelve months) (Caviedes 2010, 126). With 10,087 permits issued to the hospitality sector in 2002, representing 11 per cent of all work permits, the shortages were evident to the government (Clarke and

Salt 2003: 567). Responding to these shortages, the Labour government established a new low (or semi) skilled immigration route in 2003, the Sector Based Scheme (SBS). The SBS was modelled on the SAWS; however the SBS was targeted exclusively at immigrants wanting to work in the hotel and food-processing sectors. Like SAWS, the SBS ran on a quota system of 10,000 each year, applicants had to be aged 18–30, and successful candidates could only reside in Britain for a maximum of 12 months. While neither of these low-skilled schemes lasted—with SBS phased out from 2006 and SAWS in 2013—this was principally because of EU enlargement and the consequential replacement of low-skilled overseas labour with EU migrants. For the final six years of the SAWS, only A2 citizens (Romanian and Bulgarian) could apply, and given that citizens from Central and Eastern Europe (CEE) made up 91 per cent of applicants for SBS (Clarke and Salt 2003, 573), the closure of these schemes merely symbolised that the foreign labour that had previously filled these gaps were no longer immigrants but EU citizens exercising their right to free movement. With transitional controls lapsing for A2 migrants in January 2014, the SAWS scheme closed altogether at the end of 2013 (Home Office 2013).

Whilst the government were tinkering with existing programmes and establishing new labour immigration routes, they were also converting previously non-economic routes into working schemes. The Working Holiday Makers scheme was originally designed as a cultural exchange programme for young people from Commonwealth countries, but in 2002 the government effectively transformed this scheme into an economic route, allowing participants to switch to the work-permit scheme after 12 months. The type of work was also liberalised so that Working Holidaymakers could undertake professional and highly skilled work for the first time. Furthermore, the maximum age limit for applicants was also raised from 27 to 30. The liberalising changes made to the Working Holiday Makers Scheme, which took effect from August 2003, were in part to 'make the existing scheme as inclusive as possible of the Commonwealth as a whole', but the changes were principally made to 'provide a pool of flexible labour that can help alleviate recruitment difficulties faced by UK employers [and] help reduce the demand for labour currently supplied by illegal workers' (Home Office/IND 2002c). Since 2005 the scheme has been restricted, and the old criterion has been reinstated.

3.7.4 Student Immigration: PMI and Post-study Work Visa

Quite apart from Labour's plans on labour immigration, in a bid to generate money for the higher education sector, the government committed to doubling the number of international students studying in Britain's further and higher education institutions. In 1999, Prime Minister Blair made this pledge under the Prime Minister's Initiative for International Education (PMI1), which initially ran for six years. This commitment was then extended in April 2006 for a further five years (PMI2). The PMI scheme was established to 'secure the UK's position as a leader in international education and to sustain the growth of UK international education both in the UK and overseas' (DTZ 2011, (i)). Targets to be achieved by 2011 under PMI2 included: an additional 70,000 international (non-EU) students in UK higher education and 30,000 in further education; doubling the number of countries sending more than 10,000 students per year to the UK; demonstrable student satisfaction ratings given by international students; significant growth in the number of educational partnerships between the UK and other countries. An evaluation of PMI2 in 2011 concluded that the higher education admission target was 'likely to be met' and that 'international student satisfaction has increased on average by between 8 and 10 per cent' (DTZ 2011, 8–10). The government also removed the requirement of students to be granted permission to work part time during their studies, enabling international students to support themselves and thus making the UK 'a far more attractive destination' (UKCISA 2008, 25). Between 2004 and 2009, the total number of persons admitted to the UK for the purposes of study (including their dependants) increased by almost 60 per cent (Home Office 2010, 21). The lasting effects of this specific policy are clear, with student immigration being the largest stream into Britain between 2009 and 2012 (ONS 2012b).

For the Labour governments, student immigration was not just a lucrative market for higher education. Fee income from non-EU students rose from £672 million to more than £1,725 million between 1999 and 2004. This pool of soon to be skilled workers represented an opportunity to capitalise on the training and higher education of British institutions to fill skill shortages. In 2005 the Scottish government had secured agreement with the Home Office to establish the Fresh Talent: working for Scotland scheme, allowing non-EEA graduates from Scottish institutions to work in Scotland for up to two years after their studies (Scottish Executive

2005). The British government adopted a modified version of this programme under the Science and Engineering Graduates Scheme, which allowed graduates with science and engineering skills to remain to work in the UK for 12 months, without a job offer, and with no restrictions on the type of work. The Labour government then extended this to all graduates in 2007, easing the transitioning between visas, and allowing international students to apply for a work permit post-study under the points-based system (PBS). Dubbed 'one of the most generous schemes of its type in the world' (MAC 2009), student policy was intricately linked with the UK labour market for the first time. The government went so far as to establish a separate type of visa for such post-study work under the PBS—Tier 1 Post-study Work (PSW). The PSW visa was mutually beneficial, as it was (rightly) assumed that many international students would be pulled to the UK if they had the opportunity to work after their studies. Thus the establishment of the PSW also opportunely attracted more international students. Amongst other benefits, this new visa meant that post-study immigrants would not need to pass the resident labour market test.

3.7.5 A8 Accession

As important as these reforms were, they were all upstaged by the decision to allow citizens of the A8 countries, which joined the EU in 2004, unfettered access to the UK labour market, a decision that as we shall see had profound political consequences. A8 countries included: Czech Republic, Estonia, Hungary, Latvia, Lithuania, Poland, Slovakia and Slovenia. Malta and Cyprus also acceded but previous immigration concessions meant that there was little material change for Maltese and Cypriots immigrating to Britain. While the majority of the 15 EU member states placed restrictions on the new A8 member states, Britain (alongside Ireland and Sweden) chose not to, thus allowing citizens of the A8 countries the right to work in the UK with relatively few restrictions. The government legitimised this decision with reference to a prediction that A8 flows would be approximately 5,000–13,000 per year (Dustmann et al. 2005). However, the flows turned out to be over 20 times the upper end of this estimate, and between May 2004 and September 2005 293,000 A8 migrants entered Britain (Gilpin et al./DWP 2006, 13; OECD 2009). Poles were the dominant nationality in this migration flow, accounting for 72 per cent of all A8 migrants in 2007 (Home Office 2010, 42).

Already facing intense political pressure to 'control' Eastern European migration (to be discussed in Chap. 5), a scandal involving failures in the immigration system only served to elevate the saliency of the issue. Two whistle-blowers from the IND claimed in 2004 that there were systemic failures and widespread abuse of the migration system in Eastern Europe (Moxon 2004). A report on the scandal found that the visa system for upcoming A2 countries (Bulgaria and Romania) had been exploited (Sutton 2004). Facing severe political pressure, Immigration Minister Beverley Hughes resigned in April 2004 for misleading Parliament over her knowledge of the visa scams and the warning she had purportedly received a year earlier.

In response to the mounting political pressure induced by both the right-wing press and the visa scandal, a last minute amendment was made in February to the EU (Accession) Bill 2003; the Workers Registration Scheme (WRS). The WRS meant that any A8 migrant would have to complete 12 months of employment with one employer, before being entitled to claim any welfare benefits, and effectively acted as a data-gathering mechanism on A8 migrants. Fears over welfare shopping were unsubstantiated, with A8 employment rates reaching an annual average of 81.8 per cent in 2009 compared to 74.1 per cent for UK born workers (MAC 2009, table 3.5). The worker registration scheme closed in April 2011, but the scheme had done little in terms of any regulation of intra-EU mobility anyhow. If Labour's other liberalising reforms had not already turned Britain into a 'migration state' (Hollifield 2004), the A8 decision certainly did, with 75, 000 A8 nationals entering the UK in 2008 alone (ONS et al. 2009, 22), representing 18.2 per cent of all immigrants in the UK (OECD 2009).

3.7.6 Managed Migration: Points-based System

The Labour government conveyed this new approach to policy through the term "Managed Migration". The Labour government officially introduced the concept of managed migration in the 2002 White Paper, *Secure Borders, Safe Havens* (Home Office 2002a). Managed migration was underpinned by the belief that it was possible to encourage economically profitable immigration flows, whilst attempting to reduce unwanted immigration through increasing border surveillance. The term provided the essential framework 'for communicating a new approach by incorporating ideas regarding positive economic benefits of migration, while also maintaining a dimension of control' (Balch 2009, 622).

Three months before the 2005 General Election, the Labour government published their five-year plan for migration in their strategy paper, *Controlling Our Borders: Making Migration Work for Britain*. The focal point of the paper was the introduction of a single point-based system (PBS), which consolidated the previous 80 routes of gaining legal entry to the UK (Home Office 2005). The PBS was rolled out in phases tier by tier from 2008 and remains in place under the current government. The PBS was composed of five tiers: Tier 1, highly skilled migrants; Tier 2, skilled workers with job offers; Tier 3, low-skilled migrants; Tier 4, students; Tier 5, temporary workers and youth mobility. At the time, prospective applicants were judged and awarded points according to their human capital: qualifications, future expected earnings, sponsorship and English language skills. Whilst the system has broadly remained in tact, since 2010 the Conservative administrations have made a number of usually draconian adjustments, such as transforming Tier 1 from high skilled to "high valued".

The PBS had numerous objectives including: attracting skilled immigrants, increasing competitiveness of the highly skilled global labour pool through selective admission, creating an immigration regime susceptive to labour market needs, attracting profit and future revenue through active recruitment of international students, and enhancing tourism through simplifying entry and visa rules (Home Office 2005). The PBS was a system 'designed in a period of boom but implemented in recession' (Spencer 2011, 84). Nevertheless the system allows governments to finely tune entry criteria in consideration of changing economic circumstances and/or evidence and thus gave the Labour government (and the current government) the capacity to adapt their immigration policy in light of the 2008 global financial crisis and the recession that followed it. Alongside the PBS the Migration Advisory Committee (MAC) was established in 2007 to advise government on appropriate quotas and ways to utilise immigration to meet labour shortages (discussed further in Chap. 4).

From around the year 2000, the Labour government began to overhaul the previous policy paradigm of restriction, replacing it with an expansionary logic. Labour's second term saw the most prolific immigration reforms to date. At the same time, the government was faced with a continual barrage of criticism on asylum backlogs, administrative incompetence in the IND, ministerial scandals involving visas and inefficient border controls and consequently the government espoused a tough rhetoric and policy on these matters. But while the latter issues constantly featured in headline after headline, speech after speech, the economic immigration reforms barely faced public scrutiny or parliamentary debate; indeed this was yet to come.

3.8 2005–2010: Rowing Back

'British jobs for British workers' (Brown 2007)

The 2005 General Election best marks the juncture when the Labour administration began to "row back" on immigration. The election campaign saw the Conservative opposition talking tough on immigration, with the now infamous heading in their manifesto 'it's not racist to impose limits on immigration' (Conservative Party 2005). This caused a 'ratcheting effect' (Geddes and Tonge 2005), where 'increased salience led to increased attention that, in turn, increased the number of government statements emphasising the negativity of migratory movements' (Mulvey 2011, 1487; Saggar 1997). Immigration began to dominate the political agenda in a way not seen since the 1960s:

> Immigration was front and centre by far the biggest issue that we [Labour] talked about [in preparing for 2005 election]. At that point I would say that the negativity went in two directions. One direction was perceptions that the government were not handling the system very well. But also there was a bigger debate breaking out which was about how legitimate was the goal of a reasonably open immigration system, were there too many people coming in? And I think there was beginning to be a debate about was this having a negative effect on people at the lower end of the labour market. I would say that really the 2005 election was the break point in the sense that Labour won but I think it came away from the election recognising that immigration was and had become a very significant issue. (Interview with former SpAd, 2011)

In preparation for the 2005 election, the Labour government ran a series of focus groups led by strategy and polling adviser to the Labour Party Philip Gould and pollster to Gordon Brown Deborah Mattinson. The results, now a familiar theme, showed increasing levels of discontent with how the government were handling immigration. Already by 2006 Gould commented in his memoirs that, 'Concerns about immigration continued to heighten, and it was increasingly seen as a primary cause of other problems. People saw Labour as out of touch, not listening and gogged by sleaze and infighting' (Gould 2011, 493). Mattinson (2010, 133) similarly observed that,

> Immigration, perhaps more than any other issue, illustrates the disconnect between the voter and the Westminster Village…we described it [immigration] as a "vortex" issue, one which sucked all other issues in—the NHS is struggling? That's because it's crowded with immigrants. Can't get a job? That's because immigrants have undercut your rates…It became an issue that, in the focus groups, we always shut down and moved on from and the voters knew it.

In response to what the Labour Party perceived as rising anti-immigrant sentiment, some aspects of immigration policy and rhetoric were tightened during Blair's last two years as Prime Minister (2005–06). What is more, the foreign prisoners' scandal in 2006—when it was found that over 1,000 foreign nationals theoretically eligible for deportation had been released between 1999 and 2006, revealing major failings in senior management, and of the relationship between immigration and criminal justices systems (Hansard: Select Committee on Home Affairs 2005/2006 Fifth Report)—unsurprisingly propelled the issue up the political agenda. For example, the automatic right to apply to extend stay after four years on the HSMP was withdrawn when it was found that many on the scheme were not taking up high-skilled work. However, the courts found the retrospective rule change unlawful and ruled that the terms of the original scheme must be honoured (Spencer 2011, 92). In turn, a new, tough Home Secretary, John Reid, was appointed and with him came a boost in enforcement efforts. The new enforcement crackdown (HSMO 2007) was based on a new civil penalties regime and 'immigration crime partnerships', which incorporated other parts of the public sector regulating migration (Balch 2010, 140).

In the aftermath of the 2008 economic crisis, and with a populist backlash against immigration evident in opinion polls, Labour desperately tried to row back in its final term in office. In immigration policy, as so much else, the tone and substance of the Brown premiership were very different from the Blair government of the early to mid-2000s. Pro-immigration rhetoric was tempered and policy was tightened, as epitomised by the decision to place transitional controls on Bulgarian and Romanian citizens when their countries joined the EU in 2007 (MAC 2008). The Sector-Based Scheme was also phased out from 2006, as the quotas had not been filled. The Working Holidaymakers' Scheme was reverted back to a cultural exchange scheme, with the old conditions being fully reinstated. The criterion for a Tier 1 visa was also tightened in 2009, so that no points were awarded for a bachelor's degree or previous earnings below £20,000 (MAC 2009). Further tightening included extending advertising requirements for jobs in Tier 2, tightening the criteria for post-study work visas and a comprehensive skills review of occupations on the shortage list with a view to up-skilling existing resident or "native" workers to reduce further dependence on migrant labour (Home Office 2009).

As the economy slowed and the government's popularity plummeted, there was much less talk of the economic benefits of immigration and much more about the need to ensure a fair and robust system of controls. Indeed, the downturn seemed to have the most immediate effect on sectors of the UK economy where migrant labour was most intensely present (Scott 2008). Immigration Minister at the time, Liam Byrne, especially ramped up the tough talk on immigration, but it was arguably too late to reassure and appease public grievances.

Immigration policy innovations and administrative reforms did continue, but the former was increasingly focused on plans for strengthening border controls, including proposals for universal ID cards (Byrne 2007, 2008; Smith 2008), while the latter saw the reorganisation of the Immigration Nationality Directorate. The IND had been under much speculation and criticism over Labour's terms (and indeed Major's term in the 1990s) for general administrative incompetence, reaching a pinnacle when former Home Secretary John Reid declared the organisation 'unfit for purpose' in 2006. After a series of reviews, the IND was merged with UK Visas and HM Revenue and Customs and took on agency status to form first the Borders and Immigration Agency, which was then renamed in 2008 into the UK Border Agency (UKBA). Following reviews highlighting incompetence in the UKBA (UKBA 2011), as of April 2013, the UKBA was split into two separate units in the Home Office: a visa and immigration service and an immigration law division. In 2012 the Border Force—responsible for day-to-day operations—was separated from UKBA and became a separate law enforcement body. UKBA was then renamed to UK Visas and Immigration (UKVI) in 2013.

While the government presented tough rhetoric on immigration with an emphasis on border control, the work-related immigration reforms put in place in the early 2000s were still fundamentally intact. Furthermore, immigration flows continued to rise, with more immigrants entering Britain in Labour's final year in office (2010) than in almost any other year on record (ONS 2012a).

3.9 Conclusion

Despite the row back of the later years there is no doubt that the Labour government's record on economic immigration was one of liberalisation. Britain effectively went from being a country with a very wide conception of citizenship to a subsequent 40 years of a highly restrictive regime (with bipartisan support) and finally a system that actively endorsed economic immigration. Looked at in historical perspective, the speed and scale of the policy shift are all the more remarkable. Since post-war immigration was first regulated in the 1960s, immigration policy has evolved at a glacial pace and often in a piecemeal way. Certainly, no government had previously attempted such an ambitious overhaul of the country's immigration system and essentially there was no economic immigration policy to speak of. A myriad of factors have influenced British immigration policy over the last 50 years, primarily foreign policy and public preferences. But regulating immigration based on economic utilitarian arguments was clearly a rationale specific to the Labour governments of 1997–2010 (Balch 2010). Immigration policy was consistently made with the objective to keep foreigners living and working in Britain to an irreducible minimum. In stark contrast to Britain's historical reputation of restrictionism, the Labour governments' policy was one of expansion. The fundamental aims and objectives of immigration policy changed; hence I argue that the policy shift was akin to a third order change (Hall 1993). Yet how or why an immigration regime entirely underpinned by restriction became one of the most expansive in Western Europe in such a relatively short time span remains unclear, and it is this question to which we now turn.

References

Balch, A. (2009). Labour and Epistemic Communities: The Case of "Managed Migration" in the UK. *The British Journal of Politics and International Relations, 11*(4), 613–633.

Balch, A. (2010). *Managing Labour Migration in Europe: Ideas, Knowledge and Policy Change.* Manchester: Manchester University Press.

Bale, T., Green-Pedersen, C., Krouwel, A., Luther, K. R., & Sitter, N. (2010). If You Can't Beat Them, Join Them? Explaining Social Democratic Responses to the Challenge from the Populist Radical Right in Western Europe. *Political Studies, 58*(3), 410–426.

Blair, T. (2005). Foreword. In *Controlling Our Borders: Making Migration Work for Britain – Five Year Strategy for Asylum and Immigration.* London: Home Office.

REFERENCES

Brown, G. (2007, September 24). *Speech to Labour Conference*. Available from: http://news.bbc.co.uk/1/hi/uk_politics/7010664.stm. Accessed 13 Dec 2012.

Burns, J. (2003, July 14). Migrants "Exploited" by Shady Employers. *Financial Times*.

Byrne, L. (2007, June 19). Securing Our Identity: A 21st Century Public Good. Speech to Chatham House, London.

Byrne, L. (2008, 14 January). Border Security and Immigration: Our Deal for Delivery in 2008. Speech to Border and Immigration Agency, London.

Carter, B., Harris, C., & Joshi, S. (1987). The 1951–55 Conservative Government and the Racialization of Black Immigration. *Immigrants & Minorities, 6*(3), 335–347.

Caviedes, A. A. (2010). *Prying Open Fortress Europe: The Turn to Sectoral Labor Migration*. Lanham: Lexington books.

CEC (Commission of the European Communities). (2000). *Communication on Immigration and Asylum Policies*. COM (94) 23 Final. Brussels: Commission of the European Communities.

Clarke, J., & Salt, J. (2003). Work Permits and Foreign Labour in the UK: A Statistical Review. *Labour Market Trends, 111*(11), 563–574.

Clayton, G. (2008). *Immigration and Asylum Law*. Oxford: Oxford University Press.

Cohen, R. (1997). Shaping the Nation, Excluding the Other: The Deportation of Migrants from Britain. In J. Lucassen & L. Lucassen (Eds.), *Migration, Migration, History, History: Old Paradigms and New Perspectives* (pp. 351–375). Bern: Peter Lang.

Conservative Party. (2005). *Conservative Party General Election Manifesto: Are You Thinking What We're Thinking? It's Time for Action*. London: Conservative Party.

Craig, F. W. S. (Ed.). (1990). *British General Election Manifestos 1959–1987*. Aldershot: Dartmouth.

Dean, D. W. (1987). Coping with Colonial Immigration, the Cold War and Colonial Policy: The Labour Government and Black Communities in Great Britain 1945–51. *Immigrants & Minorities, 6*(3), 305–334.

Dobson, J., Koser, K., Mclaughlan, G., & Salt, J. (2001). *International Migration and the United Kingdom: Recent Patterns and Trends*. London: Home Office.

DTI (Department for Trade and Industry). (1998). *Our Competitive Future: Building the Knowledge Driven Economy*. London: DTI.

DTI (Department for Trade and Industry). (2000). *New Immigration Scheme*. DTI, press release: London.

DTZ. (2011). *Prime Minister's Initiative for International Education Phase 2 (PMI2): Final Evaluation Report*. London: DTZ.

Dustmann, C., Casanova, M., Fertig, M., Preston, I., & Schmidt, C. (2005). *The Impact of EU Enlargement on Migration Flows Home Office Report 25/03*. London: Home Office.

Finch, T., & Goodhart, D. (2010). Introduction. In T. Finch & D. Goodhart (Eds.), *Immigration Under Labour* (pp. 3–10). London: IPPR.
Flynn, D. (2003). *Tough as Old Boots? Asylum, Immigration and the Paradox of New Labour Policy. Immigration Rights Project*. London: Joint Council for the Welfare of Immigrants (JCWI).
Flynn, D. (2005). New Borders, New Management: The Dilemmas of Modern Immigration policies. *Ethnic and Racial Studies, 28*(3), 463–490.
Freeman, G. (1994). Commentary. In W. Cornelius, T. Tsuda, P. Martin, & J. Hollifield (Eds.), *Controlling Immigration: A Global Perspective* (pp. 297–303). Stanford: Stanford University Press.
Geddes, A. (2003). *The Politics of Migration and Immigration in Europe*. London: Sage.
Geddes, A. (2008). *Immigration and European Integration: Beyond Fortress Europe?* Manchester: Manchester University Press.
Geddes, A., & Tonge, J. (2005). *Decides: The 2005 General Election*. London: Palgrave Macmillan.
Gilpin, N., Henty, M., Lemos, S., Portes, J., & Bullen, C. (2006). *The Impact of Free Movement of Workers from Central and Eastern Europe on the UK Labour Market*. London: DWP.
Glover, S., Gott, C., Loizillon, A., Portes, J., Price, R., Spencer, S., Srinivasan, V., & Willis, C. (2001). *Migration: An Economic and Social Analysis*. London: Home Office.
Gott, C., & Johnson, K. (2002). *The Migrant Population in the UK: Fiscal Effects, RDS 77*. London: Home Office.
Gould, P. (2011). *The Unfinished Revolution: How New Labour Changed British Politics for Ever*. London: Abacus.
Hall, P. (1993). Policy Paradigms, Social Learning, and the State: The Case of Economic Policymaking in Britain. *Comparative Politics, 25*(3), 275–296.
Hampshire, J. (2005). *Citizenship and Belonging*. Hampshire: Palgrave Macmillan.
Hansard HC Deb. vol.185 cols.400-3, 7 February 1991.
Hansard HC Deb. vol.300 cols.910-911, 12 November 1997.
Hansard HC Deb. vol.299 cols.569-71, 27 October 1997.
Hansen, R. (1999). The Kenyan Asians, British Politics, and the Commonwealth Immigrants Act, 1968. *The Historical Journal, 42*(03), 809–834.
Hansen, R. (2000). *Citizenship and Immigration in Post-war Britain*. Oxford: Oxford University Press.
HMSO. (1965). *Immigration from the Commonwealth, Cm 2739*. London: HSMO.
HMSO. (2007). *Enforcing the Rules*. London: Home Office.
Hollifield, J. (2004). The Emerging Migration State. *International Migration Review, 38*(3), 885–912.
Home Office. (1998). *Fairer, Faster, Firmer- A Modern Approach to Immigration and Asylum*. London: Home Office.

Home Office. (2002a). *Secure Borders, Safe Havens.* London: Home Office.
Home Office. (2002b). *Highly Skilled Migrants Programme.* London: Home Office.
Home Office. (2002c). *Working Holidaymakers Scheme: Consultation Documents.* London: Home Office.
Home Office. (2005). *Controlling Our Borders: Making Migration Work for Britain – Five Year Strategy for Asylum and Immigration.* London: Home Office.
Home Office. (2009). *Migrant Workers Face Tougher Test to Work in the United Kingdom.* Press Release, 22 February.
Home Office. (2010). *Home Office Statistical Bulletin. Control of Immigration: Statistics United Kingdom 2009.* London: Home Office.
Home Office. (2013, September 12). Written Ministerial Statement on the Seasonal Agricultural Workers Scheme and the Food Processing Sectors Based Scheme.
Home Office, Department for Trade and Industry and Institute for Employment Studies & NOP Business. (2002). *Knowledge Migrants: The Motivations and Experiences of Professionals in the UK on Work Permits.* London: Home Office.
Huysmans, J. (2000). The European Union and the Securitization of Migration. *JCMS: Journal of Common Market Studies, 38*(5), 751–777.
Labour Party. (1997). *Labour into Power: A Framework for Partnership; New Labour: Because Britain Deserves Better.* London: Labour Party.
Layton-Henry, Z. (1992). *The Politics of Immigration.* Oxford: Blackwell.
Layton-Henry, Z. (2004). Britain: From Immigration Control to Migration Management. In W. Cornelieus, T. Tsuda, P. Martin, & J. Hollifield (Eds.), *Controlling Immigration* (pp. 297–334). Stanford: Stanford University Press.
MAC (Migration Advisory Committee). (2008). *The Labour Market Impact of Relaxing Restrictions on Employment of A2 Nationals.* London: MAC.
MAC (Migration Advisory Committee). (2009). *Analysis of the Pointed-Based System: Tier 1.* London: Migration Advisory Committee.
Mattinson, D. (2010). *Talking to a Brick Wall: How New Labour Stopped Listening to the Voter and Why We Need a New Politics.* London: Biteback Publishing.
Messina, A. (1985). Race and Party Competition in Britain: Policy Formation in the Post-Consensus Period. *Parliamentary Affairs, 38*(4), 423–436.
Messina, A. M. (1989). *Race and Party Competition in Britain.* Oxford: Oxford University Press.
Moxon, S. (2004). *The Great Immigration Scandal.* Exeter: Imprint Academic.
Mulvey, G. (2011). Immigration Under New Labour: Policy and Effects. *Journal of Ethnic and Migration Studies, 37*(9), 1477–1493.
OECD. (2009). *Sopemi Country Notes.* Available from: http://www.oecd.org/dataoecd/43/0/44068261.pdf. Accessed 01 Aug 2010.
ONS (Office National Statistics). (2010). *Long Term Migration Estimate.* Available from: www.statistics.gov. Accessed 23 Feb 2011.

ONS (Office National Statistics). (2012a). *Migration Statistics Quarterly*. Available from: http://www.ons.gov.uk/ons/publications/re-reference-tables.html?edition=tcm%3A77-256033. Accessed 17 Jan 13.

ONS (Office National Statistics). (2012b). *Migration Statistics Quarterly Report May 2012*. Available from: http://www.ons.gov.uk/ons/rel/migration1/migration-statistics-quarterly-report/may-2012/msqr.html. Accessed 19 Jan 13.

ONS (Office National Statistics), DWP (Department of Work and Pensions) and Home Office. (2009). *Migration Statistics 2008 Annual Report*. Surrey: ONS, Crown Copyright.

Paul, K. (1997). *Whitewashing Britain: Race and Citizenship in the POSTWAR Era*. Ithaca: Cornell University Press.

Prisons Reform Trust. (1984). *A Law Unto Themselves: Home Office Powers of Detention* (p. 1). London: The Trust.

Rees, T. (1982). Immigration Policies in the United Kingdom. In C. Husband (Ed.), *'Race' in Britain*. London: Hutchinson University Library.

Rex, J., & Tomlinson, S. (1979). *Colonial Immigrants in a British City: A Class Analysis*. London: Routledge and Kegan Paul.

Roche, B. (2000, September 11). UK Migration in a Global Economy. Speech to IPPR.

Saggar, S. (1992). *Race and Politics in Britain*. New York: Harvester Wheatsheaf.

Saggar, S. (1997). The Dog That Didn't Bark: Immigration, Race and the Election. In A. Geddes & J. Tonge (Eds.), *Labour's Landslide: The 1997 General Election* (pp. 147–163). Manchester: Manchester University Press.

Saggar, S. (2000). *Race and Representation: Electoral Politics and Ethnic pluralism in Britain*. Manchester: Manchester University Press.

Salt, J. (2001). *International Migration and the United Kingdom: Recent Patterns and Trends* (RDS Occasional Paper No. 75). London: Home Office.

Salt, J. (2009). *International Migration and the United Kingdom: Report of the UK SOPEMI Correspondent to the OECD*. London: University College London.

Salt, J., & Kitchling, R. T. (1990). Labour Migration and the Work Permit System in the United Kingdom. *International Migration, 28*(3), 267–294.

Scott, S. (2008). *Staff Shortages and Immigration in Agriculture (MAC Report)*. London: Migration Advisory Committee.

Scottish Executive. (2005). *Graduates – Working in Scotland Scheme*. Available from: http://www.scotland.gov.uk/News/Releases/2005/06/16102727. Accessed 23 Mar 2012.

Smith, J. J. (2008, March 6). *The National Identity Scheme*. Speech to Demos. London.

Somerville, W. (2007). *Immigration Under New Labour*. Bristol: Policy Press.

Spencer, I. R. (2002). *British Immigration Policy Since 1939: The Making of Multi-Racial Britain*. London: Routledge.

Spencer, S. (2011). *The Migration Debate*. Bristol: Policy Press.
Stevens, D. (1998). The Asylum and Immigration Act 1996: Erosion of the Right to Seek Asylum. *The Modern Law Review*, 61(2), 207–222.
Studlar, D. T. (1978). Policy Voting in Britain: The Colored Immigration issue in the 1964, 1966 and 1970 General Elections. *The American Political Science Review*, 72(1), 46–64.
Sutton, K. (2004). *Inquiry into Handling of ECAA Applications from Bulgaria and Romania*. London: Home Office.
Thatcher, M. (1978, January 27). TV Interview for Granada *World in Action* ("Rather Swamped"). Available from: http://www.margaretthatcher.org/document/103485. Accessed 1 Nov 2010.
TNA CAB 128/35, CC (61) 29, 30 May 1961.
TNA, CAB 128/43, CC 68 (13), 15 February 1968.
TNA, CAB 128/43 CC (68) 14, 22 February 1968.
TNA CAB 129/154, cp (70) 126, 31 December 1970.
TNA CAB 130/1007, Select Committee on Race Relations and Immigration: Report on Immigration, Draft Statement by Home Secretary, 1978.
TNA CAB 130/882, Official Group on Immigration Policy, Minutes from 17 September 1976.
TNA CAB 130/882'Official Group on Immigration Policy' , Minutes from 20 December 1976.
TNA CAB 164/460, Study Group on Mobility of Labour and Social Policy, Note on the Implication of Entry into Europe for Mobility of Labour, Migration and Employment, Undated.
TNA CAB 164/460, Mr Luard Quoting PM, "European Economic Community", Minutes from Hansard Commons, 16 February 1970.
TNA CAB/130/882 'Official Group on Immigration Policy', Minutes of Meeting Held on 23 September 1976.
TNA FCO 107/30, Proposed Changes to UK and Immigration Nationality Policy. Memorandum, 'Revised Immigration Rules: Essential Facts', 1979.
TNA FCO 30/3967, House of Commons Select Committee on Race Relations and Immigration: Inquiry into the Effect of the United Kingdom's Membership of the EEC on Immigration and Race Relations, Note on a Meeting held on 22 December 1977.
TNA FCO 37/1832, 'Ministerial Group on Immigration Policy', Gen 24: Ministerial Group on Immigration Policy, Meeting 2 February 1977.
TNA FCO 50/358, Negotiating Brief for UK Entry into EEC Effects on Immigration in the UK and Probable Effects on Immigration Controls If Britain Joins EEC.
TNA FCO 50/484, "Immigration Policy", Cabinet Minutes, 23rd January 1973.
TNA FCO 50/585 'Meeting on Immigration Between FCO and Home Office Ministers, Letter to Mr Cortazzi from M. Palliser, 11 June 1976.

TNA HO 344/193, 'Commonwealth Immigratin.d.on-1967', Press Notice, Undated.
TNA, HO 344/42, Letter to Minister of State, 9 November 1961.
TNA, HO 344/42, Letter to Sir Charles Cunningham on Issue of Overcrowding, 9 November 1961.
TNA HO 344/63, 'Employment: Coloured Immigrants' Contributions to British Economy', Report on Findings, no date.
TNA HO 376/130, Letter to Mr Fries from G. Meynell (Ministry of Housing), 21 December 1970.
TNA HO 376/130, Ministry of Housing and Local Government, Notes on NIESR Report on Immigrants and the Social Services, no date.
TNA Lab 8/3315, 'Economic Consequences of Immigration 1966–1970, Letter to Mr Avbery from O.F.Griffiths, 16 May 1966.
TNA T227/3216, 'Commonwealth Immigration Legislation, Brief in IC (70) 2 and IC(70)3', Letter to Mrs Case from Miss J.M Forsyth, 17 August 1970.
TNA: FCO 50/484, Letter to Prime Minister (Heath) from Brockway (President of Liberation), 14 December 1972.
TNA: HO 344/63, 'Summary of Commonwealth Immigration- New Commonwealth (Excluding UK Passport Holders', undated.
TNA: HO 344/97, Citizenship, Immigration and Integration: Opposition Green Paper, Labour Party, 1972.
TNA: T 353/68, 'Ministerial Group on Commonwealth Immigration Control, General Paper 1973, Letter to the Treasury from Sir Alec Douglas-Home, 6 March 1973.
TNA: T227/3216, 'Official Committee on Immigration and Community Relations. Proposed Immigration Legislation: ICO (70)5, 6 & 8. Brief for Meeting 8th December. Letter to Miss Forsyth from J.D Skinner, 4th December 1970.)
UK Border Agency. (2011). *Independent Chief Inspector of the UK Border Agency: Annual Report for 2010–2011*. London: Home Office.
UKCISA. (2008). *Mobility Matters, Forty Years of International Students, Forty Years of UKCISA*. London: UKCISA.
Vargas-Silva, C. (2013). *Long-term International Migration Flows to and from the UK*. Oxford: Migration Observatory.
Woodbridge, J. (2005). *Sizing the Unauthorised (Illegal) Migrant Population in the United Kingdom 2001*. London: Home Office.
Work Permits, U. K. (2002). *Review of Seasonal Agricultural Workers Scheme*. London: Work Permit UK.

CHAPTER 4

In Whose Interest? Organised Interests, Policy Networks and Collective Action

> *Yes, there was some lobbying by employers, but employers cut no ice really*
> (Interview with former Home Office Official, 2011)

4.1 INTRODUCTION

The role of organised interests influencing immigration policy change in a liberalising direction has been explored by a number of scholars (Cerna 2009). Gary Freeman's (1995, 2006) contribution is perhaps the most notable. Freeman argues that a shift to liberalisation occurs because well-organised pro-immigration clients exercise a disproportionate influence over the policymaking process compared with the unorganised, anti-immigration public. Accordingly, political elites respond to such demands to gain electoral support from interest groups. A number of scholars have applied Freeman's hypothesis to this case of policy change and have concluded that the shift in policy objectives was indeed brought about by non-governmental actors, be this businesses (Menz 2008; Caviedes 2010; Somerville and Goodman 2010), think tanks (Balch 2010) or policy networks more generally (Somerville 2007). Testing Freeman's proposition, this chapter examines whether the expansive economic immigration policies in Britain were a result of organised interest groups lobbying government

The chapter begins by discussing the multi-organisational field, differentiating types of organisations and the lobbying strategies they applied. The chapter then outlines the specific organisations that can be considered

© The Author(s) 2018
E. Consterdine, *Labour's Immigration Policy*,
https://doi.org/10.1007/978-3-319-64692-3_4

as constituting the economic immigration policy network in Britain. The chapter will then draw on the empirical evidence and present a broad chronology of the activities of non-state actors in economic immigration policymaking under Labour, where the findings reveal that the role of non-state actors in prompting policy change was limited.

4.2 Lobbying Strategies

To examine whether organised interests are able to influence and change immigration policy, we need to disaggregate the types of action organisations employ to persuade policymakers of their interest. In other words, to answer whether organised interests matter, we must understand *how* they matter. The techniques of modern lobbying, broadly defined as any attempt to communicate information to political actors (Vining et al. 2005, 151), present organised interests with a variety of tactical options to influence policy, which are dependent on a number of factors including the constellation of interests and the saliency of the issue at hand. Borrowed from Kriesi et al. (2010, 232) lobbying strategies can include the following (Table 4.1):

Despite the wide tactics available to organisations, there is a broad distinction to be made between these activities: 'inside' lobbying and 'outside' lobbying. Inside lobbying is defined as 'communication or interaction directed at policymakers or their staffs' (Kollman 1998, 35). This type of lobbying is fundamentally based on consensual activity between interest groups and policymakers and indeed 'often involves explicit collusion between lobbyists and policymakers who tend to agree with each other about some policy matter' (Kollman 1998, 23). Conversely, outside lobbying is characterised as 'appeals toward persons outside the policymaking community' (Kriesi et al. 2010, 35), which essentially involves efforts to influence the public debate, or 'going public' (Kernell 1997), with the aim of shaping policy through public pressure. In critical phases of policymaking, when for example the saliency of the issue heightens, all interest groups are more 'likely to go public than in equilibrium phases of routine policymaking' (Kriesi et al. 2010, 230).

Lobbyists provide whatever types of information will be most effective in influencing policymakers' decisions and that success frequently requires organisations to provide both electoral (outside lobbying) and policy-relevant (inside lobbying) information at different times (Caldeira et al. 2000, 53). As a result, organisations must engage in a variety of lobbying tactics, and no organisation uses exclusively inside or outside lobbying (Kollman 1998).

Table 4.1 Collective action strategies

Inside strategies	Public-related strategies/outside lobbying
Administrative strategies Serving on governmental advisory commissions or boards Participating in governmental consultation procedures Supplying information to policymakers Direct personal contact with public officials Negotiating with or informing branches of government Negotiating with or informing interest groups	*Mobilising the public, including campaigning* Making public endorsements of candidates Making financial contributions to electoral campaigns Contributing to other political campaigns Engaging n direct mail fund-raising for your organisation Organising petitions or signature collections Launching or supporting referendum campaigns Holding public assemblies and meetings Protesting or demonstrating Organising boycotts Striking
Parliamentary strategies Direct personal contact with members of Parliament or their staff Testifying in parliamentary committees or intervening in Parliament Negotiating with or informing members of Parliament	*Informing or getting informed about the public* Hiring a public relations firm to assist you in your public activities Making public speeches Running ads in the media about your position on policy issues Polling the general public on policy issues of concern to you Polling your members on policy issues
Court-related strategies Filing suit or engaging in some sort of litigation	*Media-related strategies* Writing newspaper articles Giving interviews to the media Distributing press releases Holding press conferences to announce policy positions Presenting yourself on the Web

There is a further distinction to be made between access and influence, as many organisations have access to decision makers but few have considerable influence on policy outcomes (Keefe and Ogul 1964). Maloney et al. (1994, 31) propose three levels of insider status: (1) core insider group, (2) specialist insider group, and (3) peripheral insider group. This distinction is useful, as the separation between those organisations that have influence on policy (core insider groups) can be distinguished from those groups who are only consulted on particular issues (specialist or peripheral) and may only play a consultative role.

To assess whether organised interests exerted an influence on immigration policy we need to know which organisations constituted the policy network in the period 1997–2010. We move then to a summary of the policy network including summarising the membership of the interest group and the typical lobbying strategies employed.

4.3 Britain's Immigration Policy Network

The British system of political economy has typically been regarded as a liberal market economy (Hall and Soskice 2001, 8). There is consequently a lack of coordinated wage bargaining arrangements, and firms primarily coordinate their activities via competition market arrangements (Hall and Soskice 2001, 8). As a result, there are incentives for employers to delay costly technological advancements in favour of depending on low-wage labour (Menz 2008, 156). The flexibility, availability, often superior training and educational background, and soft factors such as a stronger work ethic and commitment affiliated with migrant labour have meant that British employers on the whole embrace immigration (CIPD 2013). Declining apprenticeship and training schemes in Britain have further fostered the reliance on importing low-wage labour. Employers' preferences for foreign labour to plug shortages is however a relatively recent phenomena; in the mid-1980s for example the preferred solution for employers was to increase worker training (CBI 1984, 14). Moreover, employers have become accustomed to radical fluctuations in government immigration policy, and consequently they have safeguarded against an over reliance on importing foreign labour by concordantly pursuing other strategies to achieve labour market flexibility. Nonetheless, the Chartered Institute of Personnel and Development (CIPD) have consistently highlighted that their members are more likely to consider hiring migrant workers than the 'core jobless', such as the over-50s and lone parents (CIPD 2005, 3; TUC 2007, 12). Employers and employer associations continue to recruit international labour for a number of reasons. Yet employers rarely lobby government directly. To reach the ears of government, employers need an influential representative, usually in the form of an employer association or trade association.

4.3.1 Trade Associations/Employer Associations

The employer association that was most responsive and active on the issue of economic immigration policy under the Labour administrations was the Confederation for British Industries (CBI). The CBI is an organisation that represents 190,000 businesses in a range of sectors. The CBI is definitively a core insider group. Internal cohesion is strong and, given 'the successful monopolization' of business interests the CBI enjoys' (Menz 2011, 159), the organisation possesses a significant degree of power to influence policy.

Other large employer associations such as the British Chamber of Commerce (BCC) were also, to some extent, attentive to economic immigration policy under the Labour administrations. The BCC has its roots in local businesses (although its membership is far more diverse), representing 104,000 business members in a variety of sectors. Two other employer associations—the Institute of Directors (IoD) and the Federation for Small Businesses (FSB)—purportedly lobbied the Labour government on economic immigration policy (Menz 2008; Caviedes 2010; Somerville 2007). However, both organisations declined an interview with the author because contrarily they claim that they did not lobby on this area in the early 2000s. This being said, the IoD have expressed a 'liberal worldview on immigration' and have released statements of support for an expansionary approach to immigration policy generally, including endorsing a policy of total freedom of movement of labour within the EU (IoD 2007).

4.3.2 Professional Bodies and Sectional Interest Groups

Sectional interest groups invested in the issue of economic immigration consisted of those organisations whose sectors rely heavily on foreign labour to fill labour market shortages. Four main sectional bodies were concerned with economic immigration policy in some capacity. These included the National Farmers Union (NFU), which represents farmers and growers, Universities UK (UUK), which represents 134 universities, the British Medical Association (BMA), which is the main body representing doctors, and the British Hospitality Association (BHA), which represents hotels, restaurants and food service providers. While these organisations have insider status, they represent specialist insider groups that are consulted with on the specific migration category applicable.

4.3.3 Labour Unions

For British trade unions, immigration has long raised 'major questions related to trade union identity, purpose and roles' (Lucio and Perrett 2009, 329). In the post-war period mass labour immigration was presented by the state as temporary, which seemed to be more tolerable as this would avoid the basic fear of unions—that of losing bargaining power. However, as temporary immigration turned to settlement in Europe, and changes in the demography of migration flows (such as the feminisation of immigration) began to occur, trade unions were led to a contradictory political position:

> As far as trade unions had included migrant workers and were prepared to defend their rights, it seemed logical to back these workers in their demands and plead for lenient admission policies in such cases. On the other hand, these lenient policies would swell the number of new immigrants, some of whom would immediately or eventually enter the labour market. The dilemma of cooperation or resistance thus took a new form, and was intertwined with the dilemma of inclusion or exclusion of migrants in the trade unions. (Penninx and Roosblad 2000, 7)

There was a persistent dilemma for unions then. On the one hand, migrant labour could depress native wages and a surplus of labour could contribute to native unemployment. On the other hand, a message of international solidarity of workers and the opportunity of increasing union membership through recruiting migrant workers would give the movement stronger collective bargaining power (Penninx and Roosblad 2000, 8). Unions largely conceded to the latter strategy, and the movement has consequently become more inclusive and supportive of immigrant workers. In terms of any lobbying efforts, the focus of major unions in Britain has been on increasing migrant workers' rights and tackling irregular immigration. For example, both the TUC and Unison were involved in setting up the Gangmasters Licensing Authority (GLA), set up in 2005 to protect workers from exploitation mainly in the agriculture, shellfish and horticulture sectors. Nonetheless, unions have not lobbied for a more expansionary approach to policy as such. This is partly because 'unions for the most part, particularly with regards to EU migration were not in opposition and were in fact generally supportive [of Labour's immigration policy]' (Interview with former TUC officer, 2011).

The ideological difficulties unions has faced in relation to immigrant workers, coupled with a loss of legitimacy and power post-Thatcher, and Labour's distancing from trade unionism (symbolised by their removal of clause 4 from their constitution), have meant that unions played a less significant role in economic immigration policymaking than one might assume. This was in part political; whilst the Labour Party had once been heavily financially dependent on the trade unions (75 per cent of its funds came from unions in 1985 for example), the Party 'knew that it could not look business-friendly and independent unless it broke that link' (Brown 2011, 411). The TUC continued to have insider status under the Labour governments, but it was more akin to 'peripheral insider status', as their role was more cosmetic than instrumental.

Consistent with the literature on interest groups, core insider organisations such as the CBI predominantly used insider lobbying strategies and only used outside lobbying strategies, such as contacting the media, as a last resort in their lobbying efforts:

> We tend to do a lot more work privately—we are a lobbying organisation, not a campaigning organisation. Our campaigns are there when we need to; it will appear in the press when we need to. Most of our work we did in private and that's always the way we chose to work. We want to be constructive. (Interview with CBI, 2011)

Other organisations that had less of a central role in stakeholder engagement suggested that they may be more predisposed to using outside lobbying strategies if required: '...we're probably a bit more robust in our criticism than the CBI but then we're not the first organisation that the government turn to in the way that the CBI is' (Interview with BCC, 2012). Most lobbying activity on economic immigration policy by employer and trade associations was ultimately conducted in private: 'no employer was standing on TV saying "we need more labour migrants", it just was not something they wanted to put their heads on the parapet on, so whatever they were saying they were saying mostly privately' (Interview with Sarah Spencer 2011).

4.3.4 Think Tanks

While think tanks arguably do not lobby government in the sense that they do not necessarily call on government for specific action, they do offer their ideas to policymakers in the hope that they will be adopted (Parvin 2007, 17). Two specific non-state organisations with a promotional cause, as opposed to private interests' cause, were particularly influential and thus constitute part of the economic immigration policy network—the Institute for Public Policy Research (IPPR) and Migration Watch.

The 'practice of "thinktankery" is above all about the mediation of ideas', and the dissemination of these ideas in the public domain with the intention that this will affect policy (Schlesinger 2009, 4). Think tanks are thus in the business of knowledge production and policy discourse. Stone (1996) argues that think tanks have influence rather than direct impact on policy formulation and that they do so by 'help[ing] to provide the conceptual language, the ruling paradigms, [and] the empirical examples that become the accepted assumptions for those in charge of making policy' (Ibid, 110).

In line with the doctrines of New Public Management, some have argued that 'straddling institutions' such as think tanks, consultants and major accountancy firms have become the key advisors in government reform (Mulgan 1996). Much like their Conservative predecessors 'New Labour supporting think tanks provided a cadre of recruits for advisory posts in government and also for ministerial careers' (Schlesinger 2009, 7). The movement and sharing of staff between government and think tanks, eased by the use of secondments, 'meant that larger think tanks have become important insider groups, with the ability to push for policy reforms at the highest level of government' (Parvin 2007, 18).

When talking about the role of non-state actors in this case of policy change, it is impossible to ignore key actors from think tank IPPR. IPPR labels itself as the UK's leading progressive think tank and adopts a centre-left position. Their objectives include: combating inequality, empowering citizens, promoting social responsibility, creating a fair and sustainable economy and revitalising democracy. The close relationship between IPPR and the Labour Party began when the think tank led the ground-breaking *Commission on Social Justice* in 1992, which ignited a change and revitalisation of centre-left thinking. Described as 'Labour's civil service' (Taylor quoted in *The Observer* 2003), the close links between the Labour Party and IPPR were (and continue to be) evident. For example, 15 members of IPPR staff moved into government jobs upon Labour's 1997 victory. Critics have suggested that IPPR was part of an 'intellectualist fallacy', constrained by its special relationship with the Labour Party and the New Labour government (Bentham 2006, 167).

IPPR established itself as a 'critical friend' of Blairism (Parvin 2007, 17) and was thus a legitimised insider group under Labour. The organisation was decisively part of the policy community, particularly as it had congruent beliefs and objectives with the government:

> In the early years [2000s] we were very much on the sort of progressive wing of this....I think the economic orthodox view—you know that economic migration relatively unrestricted was a good thing for the UK economy—IPPR was a cheerleader for that. (Interview with Finch, IPPR, interview 2011)

In contrast, from its inception in 2001, Migration Watch (MW) has lobbied for a more restrictive approach to immigration policy. MW labels itself as an independent think tank, although it is arguably more akin to a pressure group (Grant 2000). While IPPR argued the case for expanding immigration policies in the early 2000s, MW called on government to reduce immigration including, but not exclusively, economic immigration:

> Our role was to provide a factual base for those who wished to criticise the government's policy. To put it another way it was ammunition. Without that, journalists who wanted to write about it didn't know where to begin. At that time the Home Office issued at least 120 pages of statistics every year, and unless you knew the system, its improved now, but then unless you knew something about the system you just couldn't remotely make sense of it, you just had to write down what the Home Office told you, which was not always true. Certainly not always the complete truth. So we were critical in that respect, nobody else was doing it. (Interview with Sir Andrew Green, MW, 2012)

Under the Labour governments MW was deemed an 'ideological outsider' (Grant 1995), as the fundamental aims of the organisation were seen as discordant with the government's objectives. Nonetheless, as founder Sir Andrew Green contends, MW's 'main impact has been on public opinion, which has been very substantial indeed. I think the concern was here anyway because real people lose their jobs and so on and so forth' (Interview with Migration Watch, 2012).

4.3.5 Charities and Humanitarian NGOs

Humanitarian interest groups are less able to influence economic immigration policy in comparison to business organisations, as the latter have a larger supporter base and superior organisational characteristics in general (Menz 2008, 91). The notable absence from the economic immigration policy network is a prominent NGO or rights-based group. While NGOs have concerns and a marginal influence on immigration policy (Statham and Geddes 2006), their objectives tend to be towards migrants' rights, especially those immigrating for humanitarian reasons. The Joint Council for the Welfare of Immigrants (JCWI) was the most prominent NGO in terms of the economic immigration agenda, but their efforts were focused on resident labour migrants' rights.

In contrast to employer associations and sectional interest groups, think tanks and pressure groups employed outside lobbying strategies far more. Groups that had no insider status, such as Migration Watch, used exclusively outside lobbying strategies, primarily media-related, such as writing newspaper articles and distributing press releases:

> The key is to build trust with the press, they need to know that you're facts are right, and we have that trust from most of the press. The Guardian have declined to mention us for the last 10 years. The Times is very pro-immigration nowadays, the Financial Times always has been for economically liberal reasons, but the rest of the press we have a strong relationship with. (Interview with Sir Andrew Green, Migration Watch, 2012)

Detecting and measuring the direct influence of think tanks on policy outcomes is notoriously difficult. Bentham (2006, 170) goes as far as to say that it is 'pointless' to attempt to do so, because of the 'complexity of disentangling cause from effect in the dense and multi-layered networks' in which think tanks operate. Moreover, think tanks produce knowledge and are 'second-hand dealers in ideas' (Desai 1994), rather than lobbyers of government as such.

Nonetheless, publishing ideas and research in the public domain is a form of outsider lobbying, and it is possible to examine the congruence between the ideas published by think tanks and the degree of concordance

Table 4.2 Economic immigration policy network under New Labour

Trade/employer associations	Professional bodies/sectoral interest groups	Labour unions	Think tanks/ pressure groups	NGOs/charities
Confederation of British Industries (CBI) British Chamber of Commerce (BCC) Institute of Directors (IoD)[a]	National Farmers Union (NFU) Universities UK (UUK) British Medical Association (BMA) British Hospitality Association (BHA)	Trade Union Congress (TUC) Unison	Institute for Public Policy Research (IPPR) Migration Watch Demos Statewatch	Migrant Rights Network Joint Council for the Welfare of Immigrants (JCWI)

[a]The IoD and the FSB declined an interview with the author

with policy outcomes. The congruence procedure devised by Yee (1996) and developed by Pautz (2011) is described as,

> …establishing congruence or concordance between ideas and the content of policy decisions. If ideas emerging from think-tanks are consistent with policy proposals from governments or leaders of political parties, this can be taken as an indication of influence'. (Pautz 2011, 190)

The congruence procedure does not claim to establish causal links between think tank ideas and policy outcomes and thus avoids the risk of post-hoc rationalisation. This being said, the extensive use of secondments under Labour, and the close links between IPPR staff and the Labour Party in particular, makes determining the influence of an organisation, rather than individual actors, problematic (Table 4.2).

4.4 Organised Interests: Power to Change?

The claim that organised interests drove the policy shift has been conflated by some with the strong economic conditions Britain was experiencing in the late 1990s. The Labour Party governed in a context of strong economic growth, low unemployment and low inflation. In 2001, unemployment reached a 25-year low of 3.5 per cent and inflation was 1.8 per cent, the lowest in Europe. As Chancellor Gordon Brown boasted in 2004,

'Britain is enjoying its longest period of sustained economic growth for more than 200 years…the longest period of sustained growth since the beginning of the Industrial Revolution' (Brown quoted in The Telegraph 2004). Employers' demands for immigrant labour are, of course, determined by economic conditions. However, whilst economic growth can trigger labour market demands for migrant labour, it is only through government decisions via policy that facilitate expansive labour immigration. As Caviedes (2010, 109) notes:

Though economic growth is a necessary pre-condition, it alone does not directly produce higher numbers of immigrants without policies that make the application process for work permits sufficiently cost- and time-effective for employers to alter their recruitment patterns.

Strong economic conditions, primarily labour market shortages and skills shortages, did however play a part in the story of policy change. Gary Freeman hypothesised that there is a 'good times/bad times dynamic in which migration is tolerated or even encouraged during expansionary phases, but becomes the focus of anxieties when unemployment rises' (Freeman 1995, 886), and in the case of immigration policy in Britain this is certainly befitting. Strong economic conditions did, to some extent, generate an increase in demand from employers for foreign labour. The demand for work permits had been rising throughout the 1990s with a sharp rise after 1994. From 1995 to 2002, total work permit applications rose steadily every year from 38,617 to 155,216, an increase of over 300 per cent. The largest annual increase was between 1999 and 2000 when applications rose by almost 42 per cent. However, this rapid increase between 1999 and 2000 was mostly attributed to the boom in the ICT sector in the late 1990s, at which point the government added ICT occupations to the shortage occupation list (the process of which is discussed below). Clearly the demand for immigrant labour was rising in the late 1990s, however only in specific sectors.

The boost in economic growth in the late 1990s also shaped employers' perception of immigration, as the movement of people was seen as bound with the new emerging global market. The business community perceived a change to how the economy was functioning in the 2000s, from a national to a global economy, which led to a fundamental change in how employers considered immigration. In the late twentieth century immigration was seen as a minor human resources issue, where work permits were sought on an ad hoc basis. But the change (or perceived change) to a globalised economy led many employers to regard immigration as an integral component of being a global firm, borne partly 'out of a conviction that the UK is now in global competition for the best brains' (Menz 2011, 13) or, as the CIPD (2004) put it, an 'international war of talent'. Immigration became tied with trade relations, a growing global workforce and investment in emerging markets. And this in turn changed both the importance of immigration for business and the way they responded to the issue in the 2000s:

> It [immigration] went from been an operational issue—am I getting service here? Am I getting the right people in? To one of—is this critical to invest in Britain as a proposition? And that was a relatively fundamental change and led to a change in the stead of what we do. (Interview with CBI, 2011)

A strong economic climate was a necessary condition for policy change. But nonetheless it would be wrong to see immigration policy change as a simple function of economic growth—a perspective that relegates the role of governments to a mere passive broker of interests. Since Commonwealth immigration began in the 1950s, the UK economy had gone through several periods of strong growth but none of these had led to policy liberalisation. On the contrary, despite several periods of growth and low unemployment, immigration policy had remained on the same restrictive path since the late 1960s. Britain's immigration system has long been a 'seasoned one, operating through a system of rules not primarily responding to labour market demands' (Salt and Kitchling 1990, 277). Thus immigration policy has not historically mapped onto economic cycles, and it has ordinarily been unresponsive to economic arguments within government (Consterdine and Hampshire 2014).

While other accounts argue that sections of the organised public, particularly employers, were the critical actors in this case of policy change (Somerville and Goodman 2010; Menz 2008; Caviedes 2010), my research found that the government had already begun to reorient the policy framework and that the organised public only had an influence in terms of tweaking the policy. Non-state actors only played this role because the Home Office made a concerted effort to become more engaged with stakeholders. Moreover, in line with the ideas of the New Labour Party, the government was generally keen on appeasing business interests; thus stakeholder engagement was government steered, not a result of interest groups lobbying.

The government had begun establishing many of the reforms in the late 1990s and early 2000s, long before interest groups became actively engaged in the issue. Programmes such as the Highly Skilled Migrants Programme (HSMP), the Innovators Scheme and the Fresh Talent Working in Scotland scheme were in development in the late 1990s with limited consultation from interest groups (HM Treasury 2000). Whilst there was informal lobbying in the late 1990s this was on a more ad hoc basis and mostly involved individual employers bargaining for more work permits with Members of Parliament or their staff, often by informal means with 'deals to score a couple of permits being done in the corridors' (Interview with Don Fynn, 2011). Employers in the 1990s were lobbying for less bureaucracy and reducing red rather than overhauling the policy framework (Interview with Roche, 2011; Interview with CBI, 2011).

At the same time, political elites such as Immigration Minister Barbara Roche had begun to articulate the government view that immigration should be seen as part of the economic growth strategy, apparently unprompted by any non-governmental influence (Interview with Roche, 2011). In 1998 the Department of Trade and Industry (DTI) had similarly cited the need to expand labour immigration routes to meet economic growth targets (DTI 1998). Essentially the Labour governments had been considering the idea of expanding economic immigration and initiating the managed migration agenda long before non-state actors became decidedly engaged on the issue.

An example of government leading autonomously on policy reform, without pressure from non-state actors, can be demonstrated with the establishment of the Sector-Based Scheme (SBS). The SBS was set up in 2003 to recruit foreign labour into the catering and hospitality sector. The sector had been experiencing labour market shortages in the late 1990s (Burns 2003; Clarke and Salt 2003, 567), and consequently one might expect this scheme to have been driven by lobbying from the hospitality industry. Yet the major trade association claimed that they never lobbied for such a scheme, rather 'the government suggested it [SBS] and we said yeah that's fine' (Interview with BHA, 2012).

The SBS was not then a product of comprehensive lobbying. It was the government that identified the shortages, and it seems it was the government, acting relatively independently, which designed and established the SBS. Ultimately labour immigration policy was elite-driven and many policy decisions went ahead despite heavy lobbying, as one former Special Advisor (SpAd) commented:

4.4 ORGANISED INTERESTS: POWER TO CHANGE?   101

One such thing was the restrictions on the Seasonal Agricultural Workers Scheme (SAWS) where I remember the lobbying from business saying there's going to be crops rotting in the fields and nothing on the shelves of the supermarket if you keep doing what you're going to do. We looked into it and thought well we don't believe that and we carried on. (Interview with former SpAd, 2012)

Both policy decisions—the establishment of the SBS in the absence of any concerted lobbying, and conversely restrictions on SAWS, despite the NFU's lobbying—demonstrate that immigration policymaking was elite-driven. As a former DTI Special Advisor commented, whilst there was a consultation approach,

...Above that you've got another layer, which, in a sense, is where policy is really made, and that (certainly in relation to immigration) was a pretty elite occupation. The policy was being driven overwhelmingly by a very small group of officials in the Home Office, some senior people in the agency plus a few key figures around the centre of government and one or two people from the Treasury. (Interview with former SpAd, 2011)

While there is limited evidence to suggest rigorous lobbying efforts from employers, skills shortages were becoming apparent to the government in the late 1990s. For example, the National Skills Task Force had identified skills shortages in the information technology, communications and technology (ITCE) sector in 1998 (Caviedes 2010, 118). Following this, in 1999 the ITCE Skills Strategy Group within the Department for Education and Employment (DfEE) published a study on ITCE skills shortages, which proposed that some occupations should be added to the shortage occupation list (DfEE 1999). Within a month, several ITCE occupations had been added to the list, and as a result, the number of work permits issued to the computing science sector rose by almost 600 per cent between 1995 and 2000 (Dobson et al. 2001, 137). The government *subsequently* set up an IT sector panel within Work Permits UK, in which representatives of the union associations Amicus and the TUC, and employers Intellect and PCG, offered information and advice as to which professions should be placed on the shortage list (Caviedes 2009).

The reforms in the ICT sector demonstrate the typical role of employers in this case of policy change; while there were skill shortages in the industry, the action to resolve these shortages was instigated by the government.

Although the changes were met with support they were not a consequence of concerted lobbying from employers.

The scarcity of resources invested by sections of the organised public, particularly employers and employer associations, further demonstrates the lack of interest from these groups in liberalising admissions. For example, a CBI interviewee dated their concern in labour immigration from around 2005. He went on to suggest that since 2005 immigration will take up 50 per cent of a policy advisor's job, whereas previously it was approximately 20 per cent (Interview with CBI, 2011). Hints at the CBI's first formal policy on labour immigration did not materialise until 2005 (Anderson 2005). Similarly, the NFU stated that they had no immigration policy until 2006 (Interview with NFU, 2011). Other organisations, such as UUK, claim that their lobbying efforts have been purely 'reactive' to government policies (Interview with UUK, 2012).

The notable exception to this scarcity of interest from non-state actors in the late 1990s was IPPR, who published *Strangers and Citizens* in 1994, edited by Sarah Spencer. This was the first substantial attempt to convince government to link immigration policy with economic policy and in turn advocated a need for evidence-based policy. Alan Findlay, one of the authors of the piece and a prominent migration economist, concluded:

> (1) that labour skill shortages could be filled by migrant labour; (2) that changes in the organisation of labour, particularly large companies require international circulation of staff; (3) that there may be grounds for entrepreneurial immigration as a device which generate jobs. (Findlay 1994, 200)

Within the report, Spencer suggested that a serious economic investigation was 'long overdue' and that 'inadequate research and a poor statistical base is a source of misinformation and weak policymaking' (Spencer 1994, 344). The report was categorically dismissed at the time by both the government and the Labour Party by Spencer's own admission. However, she claims that the now infamous 2001 report on the economics of immigration (Glover et al. 2001; discussed in Chap. 6) was set up, at the very least in part, because of *Strangers and Citizens* (Interview with Spencer 2011). It is hard to discern whether the ideas from *Strangers and Citizens* were diffused and adopted by the Labour Party over time and whether the 2001 report was a consequence of the publication as Spencer and Alex Balch (2010) suggest. A myriad of circumstances, initiatives and government motivations suggests not (discussed in Chap. 6). Nonetheless, as Alex Balch has stressed 'it is possible that elements of the Labour government had at least partially assimilated the recommendations outlined by Spencer in 1994' (Balch 2010, 148). Much of the shift in policy framing was synonymous with Spencer's recommendations, and thus there is undeniably an element of congruence between the ideas of IPPR and the policy reforms that transpired. However, it is hard to suggest that the publication of *Strangers and Citizens* (or indeed any single research publication) caused the policy shift. Rather, IPPR as an organisation personified the 'policy primeval soup' (Kingdon 1995), waiting for a 'window of opportunity' (Labour's electoral victory) to feed their policy solutions to government. IPPR thus had a prominent role in shaping, developing and reinforcing the ideas of the Labour government, often doing so by producing evidence to substantiate their beliefs.

Arguably actors from IPPR, such as Spencer, represented policy entrepreneurs in this story of policy change. The organisation persistently spoke of the positive economic benefits of immigration throughout the 1990s and early 2000s and became 'associated with the managed migration maximise the benefits/be open to the benefits of migration approach' (Interview with Spencer 2011). However, it was specific actors originally from IPPR who then moved into government positions, rather than the organisation itself, which proved to be highly influential in immigration policymaking under Labour, in particular former director of IPPR Nick Pearce. Pearce had been at IPPR until Labour's 1997 victory. He then moved to government to become David Blunkett's special advisor, first at the DfEE (1997–2001) and then at the Home Office (when Blunkett became Home Secretary). Pearce has been widely acknowledged as having a significant influence on these reforms; indeed many interviewees identified Pearce as the key actor in shaping immigration policy (Spencer 2007; Interview with Portes, 2011). Other pivotal actors on the immigration agenda, such as Matt Cavanagh (Blunkett's special advisor in the Home Office) and Sarah Spencer, similarly had careers transferring or seconding between IPPR and the Labour governments.

However, while these actors may have represented 'important interfaces for think-tank analysts trying to bring policy ideas or policy problems to ministers' attention', ultimately 'special advisors were loyal to their government, not to their former think-tank' (Pautz 2011, 204). Pearce for example commented that whilst the links between IPPR and the Labour administrations were important he did not use his work at IPPR to influence policy as a SpAd (Interview with Pearce, 2014). Indeed Pearce had done very little work on immigration in IPPR before entering government.

In contrast, the NGO sector had a very limited impact on the immigration reforms, albeit for slightly different reasons. It could be assumed that the NGO sector's limited impact in this area reflected the hierarchy of the policy network and a tendency for governments to prioritise business interests over more humanitarian concerns (Menz 2008, 91). This may be the case but NGOs had a limited interest in lobbying for expansive policies in the first instance. To the extent that NGOs were concerned with non-humanitarian migration, they were focused on the rights of resident migrants and family reunification rights rather than any liberalisation of admissions more generally. A former JCWI policy officer suggests that the shift in policy under the Labour governments somewhat confused the NGO sector, because it had been so predisposed to handling exclusively humanitarian migration:

> One of the stances that they were committed to was that they weren't dealing with economic migrants they were dealing with people whose migration was based on human rights considerations. They were adamant that they had to maintain that line of demarcation. It took quite a long time to dawn on them that at exactly the same time that you're adopting these defensive attitudes, actually the rules for economic migrants are being liberalised. (Interview with NGO 2011)

Tim Finch (IPPR, formerly Refugee Council) similarly commented that the NGO sector was ineffective in influencing labour immigration policy. Finch went on to suggest that the limited impact of NGOs may have been because there was no attempt at collaborating between the rights agenda and the private interests agenda, when the policy objectives were in many ways mutually reinforcing (Interview with Finch, IPPR, 2011).

While segments of the organised public, especially employers and trade associations, were supportive of liberalising policies, particularly initiatives to plug skill shortages (CBI 2002a), this was conveniently in line with the government's policy. Thus whilst employers input aided policy plans, it did not ignite a change in government thinking:

> I think they [employers] helped. They helped clarify in our own minds what we believed was right, namely that there were jobs that needed to be filled, there were skills that we were short of, there was economic growth that would be curtailed, there was inflation that would accelerate if we didn't do it. (Interview with David Blunkett, 2012)

Employers' demands were seen as consistent with the government's immigration control agenda. The belief was that to reduce illegality, there must be legitimate migration routes to meet the demand. Therefore whilst employers provided instrumental information as to how to regulate the system, it was only because their objectives were congruent with the governments that they had any influence:

> Yes there was some lobbying by employers, but employers cut no ice really… you've got the *Daily Mail* running this as a story—i.e. the government's losing control of immigration—on a weekly basis, there was a constant flow of attacks. The fact that the NFU wanted more Ukrainians really couldn't have mattered less politically. They wanted more people, that's true. But if it had been seen as inconsistent with the control agenda they wouldn't have cut much ice. But it was seen as consistent with the control agenda! (Interview with Home Office Official, 2011)

With the exception of IPPR, other non-state actors such as employers and unions were relatively passive on the policy issue in the 1990s and did not attempt to exert influence through any established lobbying strategies. Whilst employers, universities and most trade unions were supportive of 'the broad thrust of the government's proposals' (CBI 2002b), it was the government and the political elites that constituted it, which necessitated a 'need for action' (Kollman 1998), rather than the pro-migrant cohort of the organised public.

4.5 The Agenda-shaping Stage: Development of the Points-based System

Organised interests played a far greater and more instrumental role when the PBS was in development. Furthermore, their support for a new approach to immigration based on economic utility became more pronounced in the mid-2000s and interest groups began publicly endorsing the government's approach, thus exercising public-related strategies on economic immigration for the first time (Kernell 1997; Kriesi et al. 2010). The impetus for the increased involvement of non-state actors was, however, government steered. The Home Office set up multiple consultations on how the new system would be administered, thus encouraging administrative strategies for collective action, which then led to an increased interest and lobbying activity from sponsors (employers and education establishments) (Interview with senior civil servant 2012). In other words, the process of establishing communication with organised interests was self-reinforcing. These consultations began in approximately 2001, and as the PBS went further down the road to development, stakeholder engagement in the Home Office only intensified.

Most interviewees commented that the Home Office engagement with interest groups began from 2001 onwards, although an interviewee from the CBI suggested that Home Office engagement with stakeholders, particularly in terms of administrative strategies, became more apparent post 2005. He claimed that before 2005 the Home Office 'tended to take quite a purist view which was "we are the guardians" and actually if we slip up we get hauled over the coals by the *Daily Mail*, and therefore we will be slightly less trusting in how we work with external parties because we don't want to get ourselves into trouble'. Such concerns were purportedly 'evident in how they worked with the CBI' (Interview with CBI, 2011).

At a party political level, the increase in stakeholder engagement under the Labour governments can be attributed to the politics of the "New" Labour Party. The Labour Party were keen to distance themselves from the "Old" Labour brand associated with ideas of protectionism, nationalisation and above all Keynesian economic ideas (discussed in Chap. 5). One of the main tenets of the revitalised Labour Party was a move to become 'business-friendly', perhaps best symbolised by Prime Minister Blair labelling the Party 'the natural party of business' (Blair 1998). This was not an unmet plea from the Labour government; former CBI president Clive Thompson stated that the relationship between the CBI and the Labour government in the early 2000s was 'probably closer than at any other time in the last 25 years' (Thomson quoted in Brown 2000). The Party were keen to engage with business; hence the opening up of dialogue between non-state actors and the government was self-reinforcing and government-led.

Despite this broader shift towards the interests of business, the core tenets of "Old" Labour were to some extent maintained. The TUC continued to hold some degree of insider status, with Blair saying he 'couldn't agree more' with the TUC's stance on European immigration (Blair 2006), and the Labour administrations often legitimised immigration decisions with reference to the TUC's support on the policy, such as the A8 decision (Hansard 23 February 2004 col 23; Boswell 2009). The Labour Party's concurrent ideological pulls between "Old" Labour and "New" Labour were best exemplified by a joint position paper, published by the TUC, the CBI and the Home Office in 2005. The paper emphasised the 'joint commitment to support managed migration in the interests of the UK economy' and stressed that 'Government, employers and trade unions, in their separate spheres, are crucial' to building such a policy, with the Home Office subsequently promising 'to consult employers and trade unions about migration policies that are in the interests of Britain' (TUC 2005). Indeed Blair's consistent citing of CBI figures (PMOS 2004; Blair 2004) in relation to the economic benefits of immigration, as opposed to other government or academic figures, demonstrates Labour's appeal to the business community. By citing such figures, this in turn allowed the government to 'demonstrate a slightly broader base of democratic support' (Boswell 2009, 121) to the public than the narrower representational remit of which academics for example, in contrast, typify.

While unions' support may have been used as a legitimising device by the Labour governments, their involvement in economic immigration policy in terms of the overall framework was minimal. Although unions were supportive of many of the policy reforms, such as European enlargement (TUC 2006a), unions did not lobby government for a more liberal approach to policy. Moreover, concerns from unions, such as limited family reunion rights for temporary economic migrants, were both 'ignored and met with a complete lack of understanding on the part of government' (Menz 2008, 159). An interviewee from Unison suggested that while the union movement was starting to take a 'progressive attitude to foreign workers, [they] were still behind the government in terms of their thinking' (Interview with Unison, 2011). Nonetheless, the wider diminishing influence of unions on all types of policy partly explains their negligible impact on immigration policy. 'Consultation rather than negotiation' (Brown 2011, 407) was the bedrock of Labour's relationship with unions, as a former TUC officer argued: 'I don't think unions on any issue really under New Labour were informing policy' (Interview with former TUC, 2011). Major umbrella unions were nonetheless invited to respond to consultations and attend various stakeholder panel meetings, although there were occasions were union representatives were "accidentally" not invited (Menz 2008, 159).

The use of consultations on economic immigration policy amongst stakeholders greatly increased under the Labour administrations. The UKBA ran 24 stakeholder consultations on a variety of immigration policies, a marked increase in comparison to all previous administrations in Britain. The first major consultation on economic immigration—*Selective Admission-making migration work for Britain*—sought the views of stakeholders on the initial design of the PBS. Released in 2005, the consultation was distributed to approximately 1,602 stakeholders (IND 2005). Employers' responses in particular proved to be critical in the design of the system:

> I think we could not have been legitimately more integral [in designing the PBS]. Certainly we were fully engaged, we were closely engaged with our members, we made sure ministers and officials were well aware of where businesses were. (Interview with CBI, 2011)

While employers were not responsible for the inception of the PBS, the functionality of the system was almost entirely moulded around consultation responses (and taskforce groups) with employers and education establishments. Indeed, the 'whole question of getting a skilled based approach to it [PBS] required consultations with employer organisations' (Interview with Charles Clarke, 2012). In this sense, while non-state actors did not cause the overarching policy change, 'first and second order changes' were instigated by these organisations (Hall 1993).

The government heeded concerns from employers over the PBS causing skills shortages, and the assignment of points was adjusted accordingly. The operational policy accommodations somewhat reflected the hierarchy of the policy network. For example, concerns from the CBI over the system were notably adjusted, including producing a shortage occupation list in line with labour market expectations (CBI 2006a). Whilst there were some concerns from employers over whether the new system would lead to skills shortages, all interviewees said that their members were broadly comfortable and supportive of the PBS being ratified (Interview with BCC, 2012; CBI 2006a).

Aside from broad consultations on the operation of the PBS, the government ran a number of taskforces and forums on specific migration streams. For example, the Joint Education Taskforce (JET) was set up in 2005 by the Home Office to provide a forum for discussing student immigration. However, members commented that while the taskforce has the 'potential to impact on operational policy', the discussions within the JET 'certainly doesn't change government's mind' (Interview with UUK, 2012).

Arguably these stakeholder forums are a means for government to communicate the direction of policy to stakeholders rather than a chance for organisations to exert influence. A number of interviewees conveyed their disappointment that these forums were not a privileged lobbying opportunity (Interview with BCC, 2012). Many forums and their memberships did exhibit a symbolic or legitimising role in the sense that stakeholders were invited to hear policy announcements (although this may be privileged information) rather than a meaningful dialogue that shapes policy, indicating 'peripheral insider status' (Maloney et al. 1994). For example, the TUC were invited to most stakeholder forums, but a former policy officer contended that 'the civil service didn't see consultations with unions as a natural thing to do, even on labour market issues' (Interview with TUC, 2011).

Nonetheless, in terms of policy implementation these forums were an opportunity for interest groups, namely sponsors, to raise operational difficulties. Thus as a consequence of government inviting stakeholder input, opportunities for insider administrative action increased. But importantly, this was a useful process for stakeholders *and* the government; thus dialogue was mutually beneficial. Therefore while the 'Home Office was reasonably conscientious around consulting likely affected parties', it was still 'within the pre-determined framework and it was really about the civil service trying to understand the barriers to implementation' (Interview with former SpAd, 2011).

The emergence of a strong stakeholder dialogue on immigration policy has come about since the development and eventual establishment of the PBS. For example, global firms such as GSK suggested that their investment in economic immigration as a lobbying issue began when the PBS was established (Interview with GSK, 2012). Likewise, the BHA identified their interest in economic immigration as a lobbying issue when the PBS was introduced and stated that their lobbying and policy development had been mostly 'reactive' to government policy. They went as far as to suggest that their lobbying efforts were 'passive' before the point where the Labour government 'became agitated about the issue' (Interview with BHA, 2012). UUK cited their interest in lobbying government on student immigration when the current Coalition government introduced the annual cap on net migration in 2010 (Interview with UUK, 2012). The NFU claims that their lobbying efforts began when the government started to restrict agricultural immigration. The NFU went on to state that they had no formal policy or lobbying strategies until approximately 2007. These organisations have lobbied in response to policy changes, specifically policies that have affected the implementation and operation of immigration and government proposals that place greater liability and penalties on stakeholders (CBI 2002b) rather than lobbying for an overhaul of the policy framework.

Nonetheless, it is clear that organised interests became more publicly supportive of the PBS and the managed migration agenda more generally. In other words, many organisations that had previously shied away from employing outsider lobbying began to 'go public' and practice media-related strategies in relation to their support of the policy (Kernell 1997). The CBI in particular channelled their support in a number of press releases and statements (CBI 2006b), perhaps most notably when former President Digby Jones stressed that 'capital can't afford to be racist for lots of reasons' (BBC 2005). Jones went on to say that using controlled migration to help reduce skill gaps and stimulate economic growth 'is nothing more than common sense' (Jones quoted in The Guardian 2004). Similarly, the TUC mobilised their support for both the A8 decision and the government's general approach to immigration.

There was thus a steep increase in organisations publicly channelling their support for the new immigration system from the mid-2000s (TUC 2006a; b). This was likely in reaction to a pleading from Labour ministers. With the A8 accession approaching in May, in April 2004 Blair implored in a speech to the CBI that 'immigration and politics do not make easy bedfellows', that the reporting on immigration was 'not exactly calculated to douse the flames of concerns' and that consequently 'now is the time to make the argument for controlled migration' (Blair 2004). With Blair highlighting the sensitive politics of immigration, and outlining the roles of government, local authorities and community groups in supporting the approach to immigration policy, Blair's speech was an appeal to the CBI to publicly endorse the managed migration agenda. The government wanted employers to publicly convey their support, thereby reducing the saliency of the issue by appealing to a rationalistic business framing:

> What they [employers] wouldn't do is come out and campaign! We used to say to them "please, don't just talk to us, write to the papers, say things! Go on broadcast media, make your voices heard because that way you'll help us to carry the day". (Interview with Blunkett 2012)

The upsurge in organisations 'going public' (Kollman 1998) in supporting the policy is consistent with the literature on collective action strategies, as organisations are more likely to employ outside lobbying tactics, such as media-related strategies, when the saliency of the policy issue rises (Kriesi et al. 2010). Yet, these actions were not lobbying in the sense of trying to exert influence on government policy. Rather this was a concerted effort by the policy network to soften the public debate on immigration.

For the most part, employers and employer associations were not lobbying for an expansion of immigration policy, but were rather focused on operational aspects of the system, such as the CBI: 'Our focus has always been on the operation of the system and how business can work within the system rather than campaigning for any kind of wider liberalisation' (Interview with CBI, 2011). Patently, adjustments to operational aspects of policy can inadvertently (or intentionally) lead to an increase in admissions, such as easing the regulation and bureaucratic aspects of obtaining work visas. Nonetheless, all interviewees were adamant that they had lobbied on operational issues such as the transparency of the system and liability on employers (CBI 2002b) rather than any attempts to change the policy direction.

Prior to this period of intensified consultations in the early to mid 2000s, there were no formal stakeholder panels on economic immigration that met frequently. Communication between the Home Office and stakeholders has been a gradual and incremental process. The invitation for greater stakeholder engagement, coupled with a new system that placed greater liability on employers and education institutions, led to a steep increase in participation from non-state actors. The members of the policy network did not push for an overhaul of the immigration system. The PBS was in development regardless of interest group preferences. However, the options for how the system would operate had not been finalised, and it is at this stage that interest groups were integral.

4.6 CONSOLIDATION OF THE POLICY COMMUNITY

The period of intensified consultation, engagement and collaboration with organised interests in many ways formed the policy network that had hitherto been scattered, unorganised and lacking cohesion in policy objectives. In particular, the taskforces set up for each tier of the PBS generated greater communication between organised interests in the realm of economic immigration. This has led to a more coordinated position between non-state actors who share the same broad objectives. In other words, organisational maintenance increased, almost as a by-product of greater stakeholder engagement with the Home Office.

As has been shown, non-state actors did not prompt the overall change in the policy framework, but conversely the policy reforms made by government unwittingly generated a more formal policy network. The establishment of an economic immigration policy community was fully institutionalised by the inauguration of two new forums in Labour's last term of office: The Migration Impacts Forum (MIF) and the Migration Advisory Committee (MAC). The MAC in particular gave organised interests a systematic and explicit medium for influencing policy.

MPs Liam Byrne (former Minister for Immigration) and Phil Woolas (former Minister for Communities) launched the MIF in 2007 alongside the MAC. While the MAC was charged with looking at the economic and labour market impacts of immigration, the MIF was set up as a counter and was charged with looking at the social impacts. The MAC consists of five economists who advise the government on appropriate quotas and ways to utilise immigration to meet labour shortages based on evidence.

The MIF had a diverse membership made up of government officials and ministers, a few non-state interest groups such as the CBI and TUC, and local authorities particularly from areas where there was a high concentration of immigration such as Oldham and Peterborough. The MIF met quarterly but was deemed a failure by almost every interviewee:

> It wasn't always clear why the agendas were constructed and for what purpose... it had very little impact on policymaking. (Interview with Will Somerville, 2011)

The MIF was an attempt to integrate, in conjunction with the MAC, both the economic and social impacts of immigration, placing stakeholder input at the forefront of immigration policymaking. The forum was disbanded in 2010 because it lacked organisation, a dedicated secretariat or budget and clear objectives (Interview with former MAC secretariat, 2012).

While the MIF was perhaps a disappointment in terms of instrumental knowledge, these bodies have legitimised and formalised non-state actors' involvement in immigration policymaking. Although stakeholder engagement had emerged in the early 2000s and intensified throughout Labour's terms, it was the establishment of the MIF and in particular the MAC, and the integral role this gave to non-state actors, that changed the scope of economic immigration policymaking. While the MIF was disbanded, the MAC lives on under the current government and is viewed by ministers, civil servants and stakeholders as an essential component of immigration policymaking.

4.7 Conclusion

The organised interests approach claims that immigration policy change (at least in a liberalising direction) will be a result of non-state actors, especially employers, lobbying government for more expansive policies. Accordingly, political elites respond to such demands to gain electoral support from such interest groups (Freeman 1995, 2006). Yet the organised interests approach neglects the state as an independent actor, one with interests and agendas like any collection of actors. In this approach, the state is merely an instrument for organised interests to channel their policy preferences that 'overlooks the fact that the state—at the very least—plays an active role in defining new policy alternatives capable of securing compromise' (Boswell 2007, 79).

To assess whether organised interests matter in explaining policy change, one must examine the kinds of collective action strategies they employ and when they do so. Interest groups reacted within the terms of the policy framework and thus to suggest that organised interests drove the overall policy shift would be an overstatement. This is refuted by stakeholders' lack of interest, resources and engagement on the issue until most policy changes had transpired. Organised interests did not attempt to challenge the policy framework, and their impact on the third order change was subsequently limited.

While organised interests were not responsible for the policy shift, the policy community were broadly supportive of the reforms, which perhaps explains the absence of concerted lobbying. Non-state actors did however influence first and second order changes (i.e. the instruments settings changed but the overall goals of policy remained the same). Organised interests, especially "sponsors" in the PBS, were critical to the design and implementation of the system, but not its impetus. Nonetheless, organised interests were not the drivers of the managed migration agenda.

The incompatibility between Freeman's model specifically and the British case is perhaps because the model was originally based on the US political system, where the federal system offers multiple avenues for interest groups to lobby government. In contrast, lobbying groups in the UK encounter a fairly hostile environment, 'given the absence of corporatist structures of interest intermediation, strong executive control of the agenda, a tightly closed, highly loyal, and generally secretive civil service, and pronounced party discipline' (Menz 2008, 160). This suggests that Freeman's predictions may have greater authority in a political system perceived to be more open to the interests of big business.

Despite their minimal role in causing the policy shift, interest groups did become increasingly influential and visible (in the public domain) under Labour. However, their gaining power was a result of the governing Labour Party being keen to appease business, as well as efforts by the Home Office to engage with stakeholders, rather than pressure from interest groups to be involved. Such a stakeholder approach was reflective of a transformed so-called New Labour Party, to which we now turn as a cause for policy change.

References

Anderson, S. (2005). Migration: A Business View. In T. Pilch (Ed.), *Perspectives on Migration* (pp. 34–44). London: The Smith Institute.

Balch, A. (2010). *Managing Labour Migration in Europe: Ideas, Knowledge and Policy Change*. Manchester: Manchester University Press.

BBC. (2005, April 22). CBI Boss Rejects Immigration Cap. *BBC*. Available from: http://news.bbc.co.uk/1/hi/uk_politics/vote_2005/frontpage/4474997.stm. Accessed 11 Mar 2013.

Bentham, J. (2006). The IPPR and Demos: Think-Tanks of the New Social Democracy. *Political Quarterly, 77*(2), 166–174.

Blair, T. (1998, April 14). *Speech to American Finances*. New York.

Blair, T. (2004, April 27). *Speech to Confederation of British Industry on Migration*. London. Available from: http://www.theguardian.com/politics/2004/apr/27/immigrationpolicy.speeches. Accessed on 21 Mar 2010.

Blair, T. (2006, September 12). *Prime Minister's Address to the Trade Union Congress*. Available from: http://www.tuc.org.uk/about-tuc/congress/congress-2006/prime-ministers-address-congress-2006. Accessed on 16 Mar 2010.

REFERENCES

Boswell, C. (2007). Theorizing Migration Policy: Is There a Third Way? *International Migration Review, 41*(1), 75–100.

Boswell, C. (2009). Knowledge, Legitimation and the Politics of Risk: The Function of Research in Public Debates on Migration. *Political Studies, 57*(1), 165–186.

Brown, K. (2000, July 10). CBI Chief's Valediction is Upbeat About Euro: Industry Could Cope With the Single Currency at an Exchange Rate of DM 2.85 Sir Clive Thompson Tells Kevin Brown. *Financial Times.*

Brown, W. (2011). Industrial Relations in Britain Under New Labour, 1997–2010: A Post-Mortem. *Journal of Industrial Relations, 53*(3), 402–413.

Burns, J. (2003, July 14). Migrants "Exploited" by Shady Employers. *Financial Times.*

Caldeira, G. A., Hojnacki, M., & Wright, J. R. (2000). The Lobbying Activities of Organized Interests in Federal Judicial Nominations. *The Journal of Politics, 62*(1), 51–69.

Caviedes, A. A. (2010). *Prying Open Fortress Europe: The Turn to Sectoral Labor Migration.* Lanham: Lexington Books.

CBI (Confederation of British Industry). (1984). *CBI Annual Report 1984.* London: CBI.

CBI (Confederation of British Industry). (2002a). *Response to the Government's White Paper on Immigration, Asylum and Nationality.* London: CBI.

CBI (Confederation of British Industry). (2002b, April 7). *CBI Says Migrants Can Help UK Economy.* CBI News Release.

CBI (Confederation of British Industry). (2006a). *Consultation Response to Selective Admission: Making Migration Work for Britain.* London: CBI.

CBI (Confederation of British Industry). (2006b). *Business Summaries: Immigration and Illegal Working.* London: CBI.

Cerna, L. (2009). The Varieties of High-skilled Immigration Policies: Coalitions and Policy Outputs in Advanced Industrial Countries. *Journal of European public policy, 16*(1), 144–161.

CIPD (Chartered Institute of Personnel and Development). (2005). *Survey Report Summer/Autumn 2005.* London: CIPD.

CIPD (Chartered Institute of Personnel and Development). (2013). *The State of Migration: Employing Migrant Workers.* London: CIPD.

CIPD (Chartered Institute of Personnel and Development). (2004). *Managed Migration.* London: CIPD Publications.

Clarke, J., & Salt, J. (2003). Work Permits and Foreign Labour in the UK: A Statistical Review. *Labour Market Trends, 111*(11), 563–574.

Consterdine, E., & Hampshire, J. (2014). Immigration Policy Under New Labour: Exploring a Critical Juncture. *British Politics, 9*(3), 275–296.

Desai, R. (1994). Second Hand Dealers in Ideas: Think-tanks and Thatcherite Hegemony. *New Left Review, I*(203), 27–64.

DfEE (Department for Education and Employment). (1999). *Skills for the Information Age: Final Report from the Information Technology, Communications and Electronics Strategy Group*. Nottingham: DfEE Publications.
Dobson, J., Koser, K., Mclaughlan, G., & Salt, J. (2001). *International Migration and the United Kingdom: Recent Patterns and Trends*. London: Home Office.
DTI (Department for Trade and Industry). (1998). *Our Competitive Future: Building the Knowledge Driven Economy*. London: DTI.
Findlay, A. (1994). An Economic Audit of Contemporary Immigration. In S. Spencer (Ed.), *Strangers and Citizens* (pp. 159–202). London: Oram Press.
Freeman, G. (1995). Modes of Immigration Politics in Liberal Democratic States. *International Migration Review, 29*(4), 881–902.
Freeman, G. (2006). National Models, Policy Types, and the Politics of Immigration in Liberal Democracies. *West European Politics, 29*(2), 227–247.
Glover, S., Gott, C., Loizillon, A., Portes, J., Price, R., Spencer, S., Srinivasan, V., & Willis, C. (2001). *Migration: An Economic and Social Analysis*. London: Home Office.
Grant, W. (1995). *Pressure Groups, Politics and Democracy in Britain*. Hemel Hempstead: Philip Allen.
Grant, W. (2000). *Pressure Groups and British Politics*. Basingstoke: Macmillan Press Ltd.
Hall, P. (1993). Policy Paradigms, Social Learning, and the State: The Case of Economic Policymaking in Britain. *Comparative Politics, 25*(3), 275–296.
Hall, P., & Soskice, D. (2001). *Varieties of Capitalism: The Institutional Foundations of Comparative Analysis*. Oxford: Oxford University Press.
HM Treasury. (2000). *Productivity in the UK: The Evidence and the Government's Approach*. London: Treasury.
IND (Immigration and Nationality Directorate). (2005). *Selective Admission: Making Migration Work for Britain: Consultation Document*. London: IND.
IoD (Institute of Directors). (2007). *Immigration—The Business Perspective* (Policy Paper). London: IoD. Available from: http://www.iod.com/influencing/policy-papers/the-economy/immigration--the-business-perspective. Accessed 18 Apr 2012.
Keefe, W. J., & Ogul, M. (1964). *The American Legislative Process*. Englewood Cliffs: Prentice-Hall.
Kernell, S. (1997). *Going Public. New Strategies of Presidential Leadership*. Washington, DC: CQ Press.
Kingdon, J. (1995). *Agendas, Alternatives, and Public Policies*. New York: HarperCollins.
Kollman, K. (1998). *Outside Lobbying: Public Opinion and Interest Group Strategies*. Princeton: Princeton University Press.
Kriesi, H., Tresch, A., & Jochum, M. (2010). Going Public in the European Union: Action Repertoires of Collective Political Actors. In R. Koopmans &

P. Statham (Eds.), *The Making of a European Public Sphere* (pp. 223–245). Cambridge: Cambridge University Press.

Lucio, M. M., & Perrett, R. (2009). The Diversity and Politics of Trade Unions' Responses to Minority Ethnic and Migrant Workers: The Context of the UK. *Economic and Industrial Democracy, 30*(3), 324–347.

Maloney, W. A., Jordan, G., & McLaughlin, A. M. (1994). Interest Groups and Public Policy: The Insider/Outsider Model Revisited. *Journal of Public Policy, 14*(1), 17–38.

Menz, G. (2008). *The Political Economy of Managed Migration*. Oxford: Oxford University Press.

Menz, G. (2011). Employer Preferences for Labour Migration: Exploring "Varieties of Capitalism"- Based Contextual Conditionality in Germany and the United Kingdom. *The British Journal of Politics and International Relations, 13*(4), 534–550.

Mulgan, G. (1996). The Market Place of Ideas. In M. D. Kandiah & A. Seldon (Eds.), *Ideas and Think Tanks in Contemporary Britain* (pp. 90–107). London: Frank Cass.

Observer. (2003, September 7). Matthew Taylor Profiled. *Observer*. Available from: http://www.theguardian.com/politics/2003/sep/07/society.thinktanks. Accessed 7 Mar 2012.

Parvin, P. (2007). *Friend or Foe? Lobbying in British Democracy: A Discussion Paper*. London: Hansard Society.

Pautz, H. (2011). New Labour in Government: Think-tanks and Social Policy Reform, 1997–2001. *British Politics, 6*(2), 187–209.

Penninx, R., & Roosblad, J. (2000). Introduction. In R. Penninx & J. Roosblad (Eds.), *Trade Unions, Immigration and Immigrants in Europe 1960-1993: A Comparative Study of the Actions of Trade Unions in Seven West European Countries* (pp. 1–21). New York: Berghan Books.

PMOS (Prime Minister's Official Spokesman). (2004, February 9). *EU Accession States/Immigration*. Available from: http://downingstreetsays.com/briefings/2004/02/09. Accessed 23 Oct 2012.

Salt, J., & Kitchling, R. T. (1990). Labour Migration and the Work Permit System in the United Kingdom. *International Migration, 28*(3), 267–294.

Schlesinger, P. (2009). Creativity and the Experts: New Labour, Think Tanks, and the Policy Process. *The International Journal of Press/Politics, 14*(1), 3–20.

Somerville, W. (2007). *Immigration Under New Labour*. Bristol: Policy Press.

Somerville, W., & Goodman, S. W. (2010). The Role of Networks in the Development of UK Migration Policy. *Political Studies, 58*(5), 951–970.

Spencer, S. (1994). *Strangers and Citizens: A Positive Approach to Migrants and Refugees*. London: IPPR/Rivers Oram Press.

Spencer, S. (2007). Immigration. In A. Seldon (Ed.), *Blair's Britain 1997–2007* (pp. 341–360). Cambridge: Cambridge University Press.

Spencer, S. (2011). *The Migration Debate*. Bristol: Policy Press.
Statham, P., & Geddes, A. (2006). Elites and the "Organised Public": Who Drives British Immigration Politics and in Which Direction? *West European Politics*, 29(2), 248–269.
Stone, D. (1996). *Capturing the Political Imagination: Think Tanks and the Policy Process*. London: Frank Cass.
Telegraph. (2004, March 18). Britain Has Longest Period of Sustained Growth Since the Industrial Revolution. *The Telegraph*. Available from: http://www.telegraph.co.uk/news/uknews/1457156/Britain-has-longest-period-of-sustained-growth-since-the-Industrial-Revolution.html. Accessed 9 Nov 2011.
The Guardian. (2004, April 29). Press Review: A Whiff of Cynical Optimism. *Guardian*. Available from: http://www.theguardian.com/society/2004/apr/29/asylum.immigrationasylumandrefugees Accessed 26 May 2013.
TUC (Trades Union Congress). (2005). *Managed Migration: Working for Britain—A Joint Statement from the Home Office, CBI and TUC*. Press Release. London: TUC.
TUC (Trades Union Congress). (2006a). *General Council Statement on European Migration*. Press release. Available from: http://www.tuc.org.uk/international-issues/europe/migration/general-council-statement-european-migration. Accessed 13 Jan 2013.
TUC (Trades Union Congress). (2006b). *Making a Rights-Based Migration System Work*. Press release. Available from: http://www.tuc.org.uk/international-issues/migration/making-rights-based-migration-system-work. Accessed 13 Jan 2013.
TUC (Trades Union Congress). (2007). *The Economics of Migration*. London: TUC. Available from: http://www.tuc.org.uk/sites/default/files/extras/migration.pdf. Accessed 13 Jan 2013.
Vining, A. R., Shapiro, D. M., & Borges, B. (2005). Building the Firm's Political (Lobbying) Strategy. *Journal of Public Affairs*, 5(2), 150–175.
Yee, A. S. (1996). The Causal Effects of Ideas on Policy. *International Organization*, 50(1), 69–108.

CHAPTER 5

Do Parties Matter? Party Ideology and Party Competition

'The economics of this were sound; it was the politics that was the problem'

(Interview with Tim Finch, 2011)

5.1 Introduction

We have seen how non-governmental actors, whilst supportive of the changes, did not prompt the policy shift. The focus of this chapter is on the governing party and the elites that comprised it. The chapter contributes to the debate on whether parties matter in determining the direction of immigration policy (Schmidt 1996, 155). Judging by the established literature it would seem not (Hampshire and Bale 2015). Those who have studied the relationship between parties and immigration policy outcomes have focused their attention on party strategy on the assumption that parties are above all vote-maximisers. If we accept that 'ideology is dead' (Bell 1962), and that the art of politics is competency in 'statecraft' (Bulpitt 1996), then the focus on party strategy, rather than ideology, is understandable. But given that 'in no Western European country can politicians or political parties gain votes by favouring new immigration' (Lahav 1997, 382), expansive policies cannot be regarded as a pragmatic electoral strategy. Policies that are at odds with public preferences, such as the policy change under study here, cannot in other words be explained by electoral strategy and/or party competition. Parties do not necessarily mimic their opponents

© The Author(s) 2018
E. Consterdine, *Labour's Immigration Policy*,
https://doi.org/10.1007/978-3-319-64692-3_5

or chase the majority voter position in a 'crude Downsian fashion' in any case (Spehar et al. 2011). In short, scholars have overlooked the fact that parties operate within a certain ideological framework that somewhat conditions and constrains political action.

Following 18 years in opposition, the Party that entered government in 1997 was a very different beast from previous Labour governments. The Party had ideologically re-orientated to the centre ground, washed their hands of left-wing militant tendencies and shed protectionist policies in favour of embracing the apparently new globalised world Britain found itself in. This was said to be the 'party of change, of progress, of radicalism' (Robinson 2012, 123). This newfound third way ideology trickled down to define the agendas of many public policies (Watson and Hay 2003; Cole 1998; Harrison 2002), but it remains to be seen whether such ideas were manifested in immigration policy. The purpose of this chapter is to examine whether such changes in the party ideology shaped the immigration policy preferences of leading elites and the degree of autonomy the government had in implementing such policies.

The chapter is divided into two sections. First we look at the Labour's intra-party modernisation, particularly the shift from 'Old' Labour to 'New' Labour, and the ramifications for how this ideological modernisation changed Labour elites' preferences to immigration policy. The second section of the chapter examines the constraints on the Party's actions, both internally and externally in terms of party competition and the party political context.

5.2 Old Labour and the 'Wilderness Years'

In post-war Britain, the Labour Party was grounded in ideas of redistribution, nationalisation of services and industry and trade unionism, symbolised by the Party's commitment to Clause IV in its constitution that committed the Party to 'the common ownership of the means of production and exchange'. The Labour Party was conceived as a mass party, representing the working class people of Britain and economically the Party was committed to full employment and Keynesian economic ideals. Branded the 'loony left' by some, Labour was ideologically sympathetic to socialism. Factions of the Party were also suspicious of the EEC. Indeed, the shadow Labour Party opposed membership when the Common Market was created in 1957 (Richards and Smith 2010, 247). It was those on the Left of the Party that were especially opposed to the European project because, they claimed, membership would be a direct threat to the sovereignty of Parliament and subsequently the ability to carry out Labour's economic policy (Richards and Smith 2010, 248). There were also fundamental divisions within the Party over key issues of state intervention, fiscal policy and welfare. The historical internal divisions within the Party mean that any distinction between 'Old' Labour and 'New' Labour should be treated loosely (Driver 2011, 110).

In terms of immigration, historically the Labour Party held a bipartisan consensus with the Conservatives throughout the post-war period to maintain restrictive immigration policies. As a result 'voters perceived no real difference between the Conservative and Labour parties' approaches to immigration' (McLean 2001, 148). The bipartisan consensus that governed Britain's two major political parties since the 1970s was grounded on three pillars:

> First immigration was to be controlled and limited; second, positive measures were to be taken to ensure the integration of immigrants into the national and local community; and third, both parties were to refrain from placing issues of immigration and race at the centre of party competition. (Hansen and King 2000, 399)

In opposition, Labour objected to elements of the restrictive immigration regime, most notably the passing of the 1962 Commonwealth Immigration Act. However, on returning to office in 1964 the Labour government not only failed to repeal this Act but also passed the 1968 Commonwealth Immigration Act, which further curtailed the rights of Citizens of the UK and Colonies (CUKC) by adding an additional prerequisite for entry. This latter Act was largely established as a reactive response to Africanisation policies, which were forcing CUKCs out of the continent (discussed in Chap. 3). Liberals condemned the 1968 Act as a betrayal of Kenyan Asians fleeing persecution (Hampshire 2006, 314), and the 1974–1979 Labour government's reluctance to repeal what were seen as racist nationality and immigration laws, such as the 1971 Immigration Act, aroused much criticism from the Labour Party Race Action Group (LPRAG) (Fielding and Geddes 1998, 142). Reinforcing the bipartisan consensus that firm immigration controls underpinned good race relations, Labour focused much of their attention on tackling racial discrimination, best signified by the introduction of the 1968 Race Relations Act and the 1976 Race Relations Act.

During 18 years of opposition the Labour Party made no mention of reforming economic immigration policy. Any mention of immigration was focused on tackling the asylum crisis and developing measures to resolve racial discrimination. Documents retrieved from the Labour archives revealed that there were no debates or proposals on changing economic immigration policy in the 'wilderness years'. While Labour were emphasising the need to enhance skills for Britain's future as a 'leading industrial nation', this was framed as a need to invest in the skills of the British public, in particular educating and training young people and women entering the workplace (Smith 1990, 28; LHASC/NEC Annual Conference Report 1990; Kinnock 1991, 133, LHASC/NEC Annual Conference Report 1991). Immigration was certainly not on the political agenda for the Labour Party in the early 1990s. The only issues in relation to immigration were: to strengthen race relations, condemn the election of a British National Party (BNP) councillor in 1991, a zero tolerance approach on xenophobic or racist tendencies, advancing an education policy that better reflects multicultural society and a proposal for a campaign against racism (LHASC/LRPD (1994) Domestic and international policy committee 'resolutions from affiliated organisations: November/December 1993, PD: 3419/January 1994).

In the 1996 'blueprint' for New Labour, *The Blair Revolution: Can New Labour Deliver?*, written by New Labour architects Peter Mandleson and Roger Liddle, there was no mention of immigration and as Blair reflected 'we came to power with a fairly traditional but complacent view of immigration and asylum' (Blair 2010, 204). Likewise, in the run-up to the 1997 General Election Labour's manifesto only broached immigration in regards to abolishing the primary purpose rule and plans to reduce the asylum backlog (Labour Party 1997). Labour's strategy on immigration in the run-up to the 1997 General Election was to try to retain the post-war bipartisan consensus, by depoliticising it and preserving a fairly neutral position, reminiscent of a common centre-left party strategy to 'defuse' the issue (Bale et al. 2010; Interview with former SpAd, 2011). Economic immigration was not on the agenda for the Party, or at the very least it was not spoken about publicly.

The Labour Party's historical record on immigration—of holding a bipartisan consensus of restrictionism—demonstrates that left or centreleft parties do not necessarily implement expansive immigration policies. On the contrary, Labour's record as well as comparisons across Western

Europe (see for example Koopmans et al. 2012) illustrate that centre-left parties do not unequivocally favour expansive policies and are no less restrictive than 'Right parties on immigration control' (Givens and Luedtke 2005). A change to a centre-left governing party then cannot be considered a sufficient explanation for the policy shift. This being said, if we consider the counterfactual possibility of the Conservatives retaining office in 1997, it is very unlikely that the policy change would have ensued. Whilst the Conservative Party face pressures to liberalise policy from business interests (Bale 2008), the centre right is ideologically and traditionally committed to maintaining restrictive policies. Thus a change to a centre-left government, whilst necessary for the policy shift, cannot be said to be a sufficient condition.

5.3 New Party, New Labour

The Labour government that came to power with a landslide victory in 1997 was a very different beast from previous Labour governments. Following the heavy electoral defeat under Michael Foot in 1983, the Party's new leader, Neil Kinnock, began to shift the Party away from the hard left, a process that was continued under John Smith and then pursued with particular vigour by Tony Blair from 1994 onward. This modernisation process was formally known as Labour's policy review. Although New Labour was 'explicitly devised to demonstrate discontinuity with Labour's past' (Robinson 2012, 9), the shift in the Party from 'Old' to 'New' is best seen as an accumulation of policy strategies, reforms and ideas, as opposed to a cohesive political project created at one specific moment (Driver and Martell 2002).

Under Blair, Labour famously deleted Clause IV of its constitution, a decision of significant symbolic value; 'it was presented as a battle for the history, identity and soul of the Party' (Robinson 2012, 136). Through its 'modernisation' agenda the Party made a concerted effort to present itself as a centrist party. Central to this was an acceptance of parts of the post-Thatcher settlement. While Old Labour was committed to nationalisation, redistribution and regulation, New Labour reconciled itself to—some would say embraced—privatisation, deregulation and limits on redistribution. Labour leader Neil Kinnock's business-friendly approach stood in stark contrast to Old Labour's socialist ideals, epitomised in his keynote speech at the 1990 National Executive Committee (NEC) Conference:

There is a widespread desire to compete effectively because everybody knows that ultimately their prosperity and security depend upon it, and when companies and workforces call for change in the government's economic policies, they are not negating in special pleading, they do not want featherbedding; all they want is a context in which they can properly prove themselves. (Kinnock 1990, 129; LHASC/NEC Annual Conference Report 1990)

Blair described himself as a 'democratic socialist', but he openly admired Margaret Thatcher, and along with Chancellor Gordon Brown, accepted the need to ensure economic competitiveness through business-friendly policies. Markets were therefore 'no longer seen as subservient to governments, but as co-equals with them in a new synergetic relationship' (Freeden 2003, 46).

At the heart of this rebranding was the adoption of third-way politics, an ideological alignment described by its guru Anthony Giddens as 'a political approach that sought to reconcile economic competitiveness with social protection' (Giddens 2007). The principal values of the third way included 'Equality; protection of the vulnerable; freedom as autonomy; no rights without responsibilities; no autonomy without democracy; cosmopolitan pluralism; and philosophic conservatism' (Giddens 2003, 37). As a result, the ideological programme of New Labour included liberal, conservative and socialist elements, 'though it [was] not equidistant from them all' (Freeden 2003, 48). By culminating different traditions and philosophies those 'who attacked from either Left or Right appear to be irrational, standing against the real-world developments that have turned Left and Right into "fundamentalisms"' (Leggett 2000, 21). The third way thus provided Labour with a strong rhetorical device able to counter the Tory opposition whilst distancing themselves from their 'Old' Labour past.

Virtually every policy area was affected by the Party's ideological overhaul, including foreign policy and relations with the EU:

The pressures of domestic political competition, a change in trade union attitudes on Europe, the dynamics of the European integration process, and important changes in the Party's approach to economic policy and the role of the nation state are key elements in understanding Labour's new found Europeanism'. (Daniels 2003, 226)

From the initial phases of the Policy Review Labour's new-found Europeanism was visible. Major unions belonging to the once powerful NEC were ardently supportive of greater European integration; with the president of the Amalgamated Engineering Union and member of the TUC, Bill Jordon, going so far as to claim that the British public 'are waiting for the opportunity of Europe' (Bill Jordon, Amalgamated Engineering Union 1991, 117; LHASC/NEC Annual Conference Report 1991). The 1990 NEC Conference in particular was dominated by a need to ensure that the Party embraced the European Community calling for Britain to 'seize' Europe (Glynn Ford 1991, 183; LHASC/NEC Annual Conference Report 1991). As Blair phrased it in 1996, 'only fifteen years ago we were the anti-European party. Today, we stand for decentralising power and for constructive leadership in Europe' (Blair 1996, 259). While Blair endorsed the 'Europe ideal' support for EU integration was for him quite straightforward: 'In a world of new emerging powers, Britain needed Europe in order to exert influence and advance its interests. It wasn't complicated. It wasn't a psychiatric issue. It was a practical question of realpolitik' (Blair 2010, 533).

The Party's new Europeanism goes some way to explaining the A8 decision in 2004, as Labour's enthusiasm towards greater European integration (in comparison with their predecessors) was reflected in Britain being a champion for an enlarged EU. If Labour had maintained its 1960s Euroscepticism, it is likely that transitional controls would have been placed (Interview with Dennis MacShane, 2012). The Party shifted to a more 'defensive posture' over their time in office and couched their enthusiasm for Europe 'in terms of standing up for the national interest and defending British sovereignty' (Oppermann 2008, 171). Nonetheless, as one of the key engineers of the New Labour project said, 'It was an article of faith for the modernisers that we must be more deeply, self-confidently involved in European Affairs' (Mandelson 2010, 237). A commitment to a more positive lead in Europe was a key cornerstone separating 'Old' Labour from 'New' Labour and thus the shift in the Party made Labour's Eurosceptic days look like a distant memory.

Notwithstanding how and why Labour transformed, it is clear that the Party modernised their political image and policies. Elements of 'Old' Labour remained intact but were renovated, such as a notion of social democracy; from a humanitarian stance of a 'right to work', to an emphasis on duties, obligations and citizenship, expressed as a 'duty to work'. Nonetheless, the pillars of Old Labour were seen to be out-dated and inconsistent with modern political economy:

> The New Right argued that the state was doing too much: high public expenditure, especially on welfare, and high taxes were undermining incentives,

crowding out public investment and creating a dependency culture. Social democracy, whether practised by the Tories or by Labour, was the villain leading Britain, if not down the road to serfdom, then at least down the drain. (Martell and Driver 2006, 31)

5.4　THE THIRD WAY AND IMMIGRATION

The modernisation of the Party is essential to understanding why the Labour government liberalised immigration policy in the early 2000s and also why previous governments—both Labour and Conservative—had not. The policy change of the early 2000s was effectively the application of New Labour's governing philosophy to immigration policy. The ideological reorientation of the Party is critical to understanding why economic immigration policy changed because of three contingent elements of Labour's new found philosophy: the Party's neoliberal economic programme, including counter-inflationary measures and labour market flexibility, and underpinning this programme, the Party's diagnosis of the global political economy. This economistic reasoning was reinforced by Labour's culturally cosmopolitan notion of citizenship and integration. Coupled with Labour's historical values of openness and tolerance, an expansive immigration policy 'made political sense' (Interview with Jon Cruddas, 2011).

Rejuvenating the Party fundamentally required presenting an appealing package of ideas that were fresh, current and above all else appealed to the British electorate. Throwing out Old Labour socialist ideals was not enough. The Party needed to articulate and deliver an original and progressive policy programme. Besides the literal rebranding of the Labour Party as 'New' Labour, the term that the Party chose to communicate their reinvigorated ideology was modernisation. Blair persistently spoke of the need for modernisation; indeed he used the term no less than 21 times in one speech at the 1997 Labour Party Conference. Modernisation was originally devised to signify that the Labour Party was 'throwing off historical obsessions and providing a new sense of direction' (Taylor 1997, 104–5), but it quickly became the buzzword to describe all planned policy reforms. Having denounced the Keynesian demand management, state intervention, 'Old' Labour approach to monetary management, New Labour claimed that the new structural reality was one of increasingly non-negotiable economic imperatives, frictionless markets that needed to be harnessed rather than resisted. The country was said to be structurally dependent on capital, and consequently what was required was minimum state intervention and a modernised approach to public policy. In stark juxtaposition to Old Labour, good government was said to be minimal government, and injustices of capitalism could best be resolved by working within the market rather than against it (Richards and Smith 2010, 244). Central to this was an appeal to the need for counter-inflationary credibility and in turn a labour market policy that above all required flexibility and competitiveness.

The critical objective of Labour's economic policy was stability; as Chancellor Brown proclaimed, 'stability is the necessary precondition for all we do' (Brown 2001, 44). Economic mismanagement was widely blamed for Labour's electoral loss in 1979, and consequently economic incompetence dogged the Party throughout the 1980s and early 1990s. To achieve a sustainable framework for economic stability, the Labour governments focused their efforts on counter-inflationary measures. Since the 1970s, the general theme of macroeconomic theory has focused on the presumed inflationary bias, which occurs when elected officials regulate monetary policy decisions (Watson and Hay 2003, 294). Indeed, the partial victory of the neo-liberal approach is by virtue of its focus and prioritisation of controlling inflation (Cerny and Evans 2004, 53). Of critical importance for the Labour Party then was the capacity to formulate and enforce suitable mechanisms (such as independent central banks authorised to deliver price stability) that would consign those responsible to a tough counter-inflationary policy. The objective to achieve counter-inflationary measures was imperative to Labour's economic policy.

To achieve low inflation, Labour prioritised labour market flexibility. This was built on the notion that supply increases demand, and thus extra competition in the labour market will reduce inflationary pressures (Glyn and Wood 2001, 55). Labour market flexibility was at the heart of Blair's vision of the competition state and represents the bedrock of Labour's economic policy both independently and in terms of its effect on keeping inflation down.

It was the Party's emphasis on labour market flexibility that some interviewees suggest was the core reason why economic immigration policy was liberalised:

> I think it [immigration policy] was basically a product of its broader economic framework which was one of economic liberalisation.... The question was what were the features that could stop it, and one of them was perceived

to be labour market inflexibility, wage inflexibility primarily and the demand for labour. And that's basically where the migration policy fitted into it..... I think it was consistent with the overarching modernisation story of economic modernisation and modernisation of civil society. It worked alongside its cosmopolitanism. So national boundaries or a patriotic story didn't necessarily fit into that. (Interview with Jon Cruddas, 2011)

If we assume that Labour's economic policy was of paramount importance to the Party, and that labour market flexibility was key to achieving this policy, an expansive economic immigration policy dovetailed with this objective. The mantra of attracting the 'brightest and the best' was synonymous with the rhetoric of competiveness used to legitimise Labour's orthodox macroeconomic policy, as explicitly outlined by Immigration Minister Barbara Roche in 2000:

> As with other aspects of globalisation, there are potentially huge economic benefits for Britain if it is able to adapt to the new environment. We are in competition for the brightest and best talents—the entrepreneurs, the scientists, the high technology specialists who make the global economy tick. In order to seize the opportunities of the knowledge economy, and to play a constructive part in shaping these huge changes, we need to explore carefully their implications for immigration policy. (Roche 2000)

Britain's out-dated and insular approach to migration was, so the argument went, entirely unsuited to the needs of a knowledge economy in a globalised world. This position was articulated in Labour's 2001 manifesto, which explicitly linked changes in the economy with a need to modify immigration policy: 'As our economy changes and expands, so our rules on immigration need to reflect the need to meet skills shortages' (Labour Party 2001, 34).

But fundamentally underpinning Labours' economic programme, and indeed the inception of 'New' Labour altogether, was an uncompromising belief in globalisation. The business school globalisation thesis propagated throughout the 1990s offered the Party a way of rhetorically imploring the need to ensure competiveness. The globalisation thesis presumed that 'in the face of global competition, capitals are increasingly constrained to compete on the world market' (Cole 1998, 315). Thus the 'argument is that these capitals can only do this in so far as they become multinational corporations and operate on a world scale, outside the confines of the nation state' (Cole 1998, 315). By diagnosing globalisation as inevitable and an unavoidable exogenous condition, the Party could offer appealing solutions to what they insisted were the new economic realities. In constructing a vision of an inevitably globalised economy, the Party rendered Labour's economic policy an absolute necessity to Britain's future prosperity, thus strategically blurring the lines between the inevitable and the desirable (Watson and Hay 2003).

Economic globalisation was presented as a non-negotiable, external constraint, an irreversible fact of life and a natural development of capitalism that could not be controlled by human agency. According to Blair the only 'rational response' to globalisation was 'to manage it, prepare for it, and roll with it' (Blair 2006). And Brown likewise proclaimed that 'we cannot any longer escape the consequences of our interdependence. The old distinction between "over there" and "over here" does not make sense of this interdependent world' (Brown 2007). By depicting such a phenomenon as inevitable, Labour's policy solutions looked apt to respond to such forces and thus desirable. The rhetoric espoused followed a simple logic: that globalisation is inexorable, it demands a strict macroeconomic orthodox approach to economic policy (Balls 1998), and that only Labour can reap the benefits of this new globalised economy:

> With new Labour, Britain can seize the opportunities of globalisation, creating jobs and prosperity for people up and down the country. The choice is to go forward to economic stability, rising prosperity and wider opportunities with new Labour. Or go back to the bad old days of Tory cuts, insecurity and instability. (Labour Party 2005, 29)

Faced with the allegedly uncontrollable exogenous constraint of globalisation, which accordingly 'has changed the nature of power held by nation states' (Blunkett 2002), Labour claimed there is simply no alternative (Watson and Hay 2003).

The diagnosis or strategic blurring of globalisation as both inevitable and desirable is in many ways the defining reason why immigration policy shifted under Labour. After all, according to the government 'Migration is driven by globalisation' (Home Office 2005, 11; Blair 2004), and thus states must respond to the imminent flows of people and make it work in their favour. However, the Party did not necessarily explicitly or consciously make the link between the rhetoric of globalisation on the one hand and expansive immigration policies on the other. Rather, the logical extension of an ideology that hinged on globalisation was so deeply entrenched that with immigration being 'the human element of globalisation' (OECD 2009), it was likewise assumed to be both inevitable and intrinsically positive by the leading faction of the Party. Tantamount to globalisation, immigration was deemed uncontrollable and therefore something that needed to be harnessed:

> Migration is a feature of globalisation in the same way that capital moves, so does labour moves across boundaries. And every attempt to close one door simply means that people move into another... I said [to constituents] "I can't stop the change, the change is happening and that's because I believe it—migration—to be part of globalisation, but what I can do is try and make it work for you". (Interview with Margaret Hodge, 2012)

> I worked under both Blair and Brown and other ministers and they shared a basic view which was that economic immigration was both good for the economy and inevitable, and sort of an inevitable feature of globalisation which was something that needed to be harnessed. These things were good for the economy overall and it was important for the country to not close its mind. (Interview with former SpAd, 2012)

> The main thing is that the reorientation on economic policy of the centre left—away from Keynesian demand management towards a more explicit embrace of globalisation—lent itself more firmly towards embracing immigration too. The emphasis on skills and education and openness to global markets meant that you had people more open to arguments about migration being an important component of a successful economy. (Interview with former SpAd, 2014)

With a governing Party committed to a programme of open markets, competiveness, flexibility and an 'open global society' (Blair 2000a), it would seem at odds to have anything but an expansive immigration policy. As a result, the expansive reforms became 'an unquestioned policy by the early 2000s' (Interview with former SpAd, 2012).

This is not to suggest that elites passively responded to global forces and thus expanded immigration, as has been implied by other accounts that focus on organised interests and economic conditions (Menz 2008; Interview with Portes, 2011). This would be to neglect the agency of the very actors who devised the policies. On the contrary, Labour deployed such a powerful discourse that globalisation became an accepted reality of governance across Westminster (Watson and Hay 2003). The argument here is that the logic of there being no alternative to globalisation permeated and trickled down to demarcating the objectives of many public policies, including immigration. The powerful and persuasive ideology of globalisation, devised and deployed by the Party, took on an unquestioned logic, an idea to define and shape policies around. Thus by appealing to the forces of globalisation as an overarching ideology which any government must accommodate, the Labour Party could claim that policy was 'depoliticised' (Burnham 2001), simply reacting to exogenous uncontrollable forces. And this logic was applied—it seems at least initially unconsciously—to immigration policy.

The conclusion that there was no alternative to globalisation was a political construction devised by the Party. Yet this ideology held such a powerful force and came to dominate and shape what Labour did to such an extent that leading Labour elites guided their political action according to the logic of globalisation and extended the rationale to other areas of policy which were traditionally influenced and made in reaction to public preferences. The irony is that in building and adhering to a party ideology that was modern and appealing to the British public, the Party wound up having a public policy that was fundamentally at odds with public preferences (Saggar 2003). It is in this sense that the policy change was an unintended outcome of intended action.

5.5 Cosmopolitan Pluralism

This economistic reasoning was reinforced by a secondary aspect of the Third Way, its cosmopolitan pluralism, which was part of New Labour's wider project of fostering a progressive interpretation of British identity based on ideas of tolerance, openness and internationalism, 'enlightened patriotism' as Blair called it (Driver and Martell 2002, 144; Shi 2008). Labour capitalised on (or arguably made) the ideas of the time, and fashioned a policy for the 'winners of globalisation', that is the 'people who benefit from new opportunities resulting from globalisation, and whose life chances are enhanced' (Kriesi et al. 2008, 5). With the cultural cleavage transforming—polarising a cosmopolitan, pro-European position on the one hand, with nationalism, isolationism and protectionism on the other—Labour appealed to the former in a bid to become a centrist party:

> In becoming more pro-European Labour shifted away from the views of its traditional support base in the working class and towards the target group

of the educated middle classes that provided the core of new party members and a basis for the Party's electoral success in the 1990s. (Kriesi and Frey 2008, 196)

In terms of how party ideology shapes elite preferences towards immigration, factions of the Party viewed immigration as an intrinsically positive thing which was consistent with Labour values, especially the longstanding emphasis on equality of opportunity for ethnic minorities:

> Most of the establishment [Labour Party] believed in it—that immigration was socially and intrinsically a good thing..... And I think given how important race and multiculturalism had become within the Labour party, the impression I've had is that most people in the Party, most people on the left, most liberals, tend to think of immigration as an intrinsically good thing.... there are particular policy levers but I think they're informed by a set of political, cultural assumptions. (Interview with former SpAd, 2012)

Numerous interviewees commented that there was an assumption in the Labour Party that embracing immigration was a logical reflection of Labour values: 'It [migration] is at the heart of Labour values—tolerance, openness, equality—all those sorts of things. I think they're very closely linked to celebrating migration' (Interview with Margaret Hodge, 2012). Labour MP John Denham suggested that the Party took 'a fairly *Guardian* type view of these issues', to the extent that the Party perceived immigration as a fundamentally positive thing (Interview with John Denham, 2012).

With values of openness and tolerance inscribed at the heart of the Party, Labour's approach to integration complemented the notion of having an expanding immigrant population. The Party embraced diversity and their integration policy was grounded in multicultural ideas. This was also reflected in their concept of British national identity, based on 'shared values not unchanging institutions' (Blair 2000b). Blair's notion of British identity was founded on ideas of liberal pluralism, communities, inclusivity, openness, tolerance and rights with responsibilities (Blair 2000b; Martell and Driver 2006). Linked to embracing diversity was the idea that migration was part of a wider framework around civic republicanism. David Blunkett in particular advocated an approach that saw migration and civic republicanism as two sides of the same coin. In a pamphlet written for IPPR, Blunkett suggested:

We have not historically had a strong tradition of common citizenship, or what is termed civic republicanism in political philosophy. We have tended to the laissez-faire, resting on anti- discrimination legislation to tackle racism and exclusion, and loose association in the public realm, rather than giving real meaning and content to the acquisition of citizenship. (Blunkett 2003, 15)

Blunkett was attempting to redefine the meaning of British citizenship, shifting the focus to 'active citizenship' and in turn linking the idea of duties and entitlements to migration. The ID card scheme, for example, was envisaged as 'a sort of badged membership of the political community' (Interview with Don Flynn, 2012). A flexible comprehension of national identity that was not based on any kind of ethno-cultural or even historical notion of citizenship could contain and respond to an expanding immigrant population.

The alignment of economic liberalism with national-cultural progressivism also helps to explain why previous governments, in particular the Conservative administrations of the 1980s and 1990s, had *not* previously liberalised immigration policy, for New Labour's economic liberalism was certainly not new. Indeed, some argue that the Labour Party's adaptation was simply an 'accommodation' of changing conditions and 'catching up' with the neoliberal doctrine of Thatcherism (Heffernan 2000). The Party's modernisation undoubtedly resulted in a shift to New Right politics in terms of their economic policy (Hall 2003). But the Thatcher governments had stopped well short of a liberal approach to immigration. This was principally because Thatcher's commitment to the free economy was counterbalanced (some would say contradicted) by conservative nationalist values. As Giddens observed of Thatcherism, 'devotion to the free market on the one hand, and to the traditional family and nation on the other, is self-contradictory. Individualism and choice are supposed to stop abruptly at the boundaries of the family and national identity, where tradition must stand intact' (Giddens 2003, 34).

Whilst Thatcher's economic programme logically implied a more expansive immigration policy, the embedded commitment to national identity and tradition in the Party's ideology pulled in the opposite direction: '[she] adopted a populist approach which depicted immigration as a threat to British nationhood…Thatcherism exhibited a cultural integrationist perspective which held that successful integration required that ethnic minorities accept the primacy of British values' (Lynch 2000, 62). In the event, the latter won out as the Thatcher and Major governments maintained and in the former case even extended the status quo of immigration restriction, in both rhetoric (e.g. Thatcher's infamous 'swamping' statement in 1978) and legislation (all of the major Acts of the 1980s—the 1981 British Nationality Act, the 1987 Immigration (Carrier's Liability) Act and the 1988 Immigration Act—had a restrictive intent). From the perspective of immigration politics, what was 'new' about New Labour was its embrace of neo-Thatcherite ideas, adapted and extended into a hyper-globalist understanding of the economy, and yoked not to an insular account of national identity but to an inclusive and even cosmopolitan conception (see e.g. Blair 2000b; Consterdine and Hampshire 2014).

A new centre-left governing party and its reorientation make economic immigration policy change seem a natural consequence of New Labour's ideology for three interconnected reasons: an uncompromising belief in globalisation, a need for labour market flexibility and Labour's historical values of openness and embracing diversity dovetailing with a flexible, cosmopolitan pluralist vision of national identity. For all these reasons, economic immigration policy mirrored New Labour ideology.

This being said, ideology works only as a broad infrastructure to guide action. Few office-seeking parties prioritise ideological purity over electoral expediency. Parties are office-seeking organisations, and thus they develop and change policies in line with what they perceive to be public preferences in order to capture voters. Strategy is essential for party success and this requires a trade-off between ideological integrity and electoral calculation. It is the other side of immigration policy (asylum) that provided this trade-off.

5.6 Control and Continuity: The Other Side of Immigration Policy

It may have made ideological sense to pursue an expansive economic immigration policy, yet asylum policy ran in the opposite direction. By being firm on asylum both in rhetoric and policy the Labour government were attempting to appease electoral concerns whilst pursuing an expansive economic immigration policy, albeit somewhat behind Whitehall doors. The two streams were seen as interconnected, as Immigration Minister Beverley Hughes maintained: 'the strategies of bearing down strongly on abuse of the asylum system and opening up legitimate routes in, are the two essential strands of a coherent policy' (Hansard 28 October 2002, col 28).

5.6 CONTROL AND CONTINUITY: THE OTHER SIDE OF IMMIGRATION POLICY

The bifurcation between 'wanted' and 'unwanted' immigration allowed the Labour government to be tough on asylum whilst almost tacitly pursuing expansive economic immigration. The two strands of policy can thus be seen as a complementary strategy for the Labour governments. As Mulvey puts it, 'Policy was control on one side (asylum) to allow continuity on the other (labour migration)' (Mulvey 2011, 1478). The leading faction of the Party purportedly believed that if they could resolve the asylum backlog and 'unwanted' immigrants, the public would be supportive of expanding 'wanted' immigrants:

> They thought that if you could deal with the sense of a problem around the large number of asylum claims which were thought to be unfounded, and issues around people coming to claim benefits, if you deal with those things then what you're left with is the masses of the people who are genuinely coming to work or to study. They were convinced that in that case the great majority of the public's concern about this would drop away, and I think there's an argument for saying that they were mistaken about that. (Interview with former SpAd, 2012.)

> On asylum they were sort of trying to be quite tough and restrictive, meanwhile they were running, in retrospect what looked like positively reckless open policy of economic migration – reckless, not with the economy, not with real things if you like, reckless in terms of their political reputation. (Interview with Tim Finch, 2011)

The White Paper *Secure Borders, Safe Havens* (Home Office 2002), written by Home Secretary David Blunkett and his special advisor Nick Pearce, was an explicit attempt to articulate this dual strategy to the public, premised on the Blunkett equation, outlined in Chapter. 3 (Balch 2010). The purpose of the paper was to provide an integrated statement that drew together the two different threads (asylum and economic). Blunkett's view was that,

> the big spikes in asylum were partly about economic migration. We needed to reform the system and the Prime Minister was firmly of that view. David felt that if there were managed migration routes to meet labour market needs then people could use those so they wouldn't chose to go through the asylum route and therefore you would have a diminution of asylum claims, and that this would also alleviate irregular migration. So his whole purpose was to bring it into an integrated and legal framework. He hoped to secure

public confidence in doing this. He wanted to the British people to understand that they could have confidence and trust in the systems the government was running. (Interview with Nick Pearce, 2014)

Arguably the dual strategy did not succeed in appeasing public concerns; immigration rose as an issue of voting importance throughout Labour's time in office, becoming a top three voting issue by 2003 and preferences pointing fairly unequivocally towards restriction (Ipsos Mori 2004; Saggar 2003). Many interviewees suggested that the Labour government's failing was that the policy was not effectively communicated to the public, or in other words 'they didn't sell it enough, they didn't work hard enough to convince the public that this was a good thing' (Interview with Tim Finch, 2011).

5.7 Constraints on Party Action: Internal Dissent and External Competition

Whilst the ideological repositioning of the Labour Party proved to be an important component in explaining the policy shift, governing parties do not operate completely autonomously. Governing parties can and do face constraints both internally from within factions of their party and externally from opposition parties seeking to gain office. Internally, governing parties can face constraints in implementing policy from its parliamentary party, in this case the Parliamentary Labour Party (PLP). It is rare to see a government defeated in the Commons by its own MPs, but nonetheless MPs can and do collectively rebel and in turn block policy. After all, party unity in the parliamentary party is a key component of the art of statecraft (Bulpitt 1996). We now turn to whether the Labour governments faced such constraints in implementing the policy reforms by first addressing resistance from within the PLP and subsequently examining the party political context and the extent to which opposition parties challenged the reforms.

5.7.1 Dissent and Intra-party Conflict

After losing office in 1979, the Labour Party was 'racked by internal disputes over the Party's future direction' (Fielding and Geddes 1998, 66). Factionalism, rebellion and internal dissent were frequent problems for the Party throughout the post-war period (Cowley 2002). This was a significant problem for the Labour Party given that party parliamentarians are one of its most public manifestations and therefore 'parties that want to appear united want their MPs to appear united' (Cowley and Stuart 2003, 316).

For the reinvigorated 'New' Labour Party, cohesion and consensus were a key objective, and to achieve this strong leadership was required (Shaw 2002). The Party leadership of both Blair and Brown were known to go against their own Party at times. Blair had a 'high belief in his ability to control events' (Dyson 2006, 303) and scholars have suggested his time in government was a 'command premiership' where he strove to be 'the ruler of the state rather than its servant' (Hennessy 2005, 15). Blair's strategy was to talk up how bad divisions had been previously in order to talk about how good things had become in the Party (Bale 1999). This allowed the 'Party's supposed cohesion to be presented as an important part of what made it new, and (just as importantly) what made Labour different from the Conservatives, [who were] then widely seen by the public as split' (Cowley and Stuart 2003, 316).

Certainly in terms of the number of revolts, the PLP conformed to Blair's wishes, with the first term having the fewest rebellions by government backbenchers since 1955 (Cowley and Stuart 2003, 317). In turn, the PLP in Labour's first term was criticised for being 'timid, sycophantic, acquiescent and cowardly' (Cowley 2005, 3). Most previous Labour leaders were accused of not being in control of the Party, but 'after 1997 the complaint was that that the leadership was too much in control' (Cowley 2005, 3). Cowley and Stuart argue that the changes in the composition of the PLP made rebellion less likely and thus acted in the interests of the party leadership (Cowley and Stuart 2003, 324). On all measurable criteria associated with rebellion, the parliamentary party fell short: 'there were fewer left-wingers; there were probably more delegates; there were more young MPs; and there was little organised opposition to the government' (Cowley and Stuart 2003, 324). As a result, it is fair to say that the first term of office was a consensual time for the PLP, and therefore for the Labour government.

This is not to suggest that rebellions did not occur however; there were deep divisions and rebellions on Labour's asylum bills. In particular the 1999 Immigration and Asylum Bill saw 61 Labour members sign an Early Day Motion put down by Diane Abbott, focusing on the issue of the children of asylum seekers and urging the government to reconsider their proposals (Cowley 2005, 67). Chair of the All-Party Group on Refugees,

Labour MP Neil Gerrard, then moved an amendment to remove families with children from the voucher system altogether. However, Home Secretary Jack Straw made concessions and with some persuasion Straw 'chipped away at the would-be rebels until Gerrard, seeing how little support he now had as a result of Straw's concessions, decided not to press his amendment to a vote' (Cowley 2005, 69). Nonetheless, 17 MPs voted against the whip on the 1999 Immigration and Asylum Bill and the divisions over the somewhat controversial measures—namely the voucher system that was soon scrapped—was certainly evident in the PLP. But this remained a 'narrow interest', and aside from controversies over Labour's asylum plans, the government faced no opposition from the PLP in terms of their plans on economic immigration.

This being said, the lack of division on the changes may have been because the majority of reforms required no primary legislation. No legislation equalled no vote, and therefore no rows. The policy reforms were formulated by the leading Labour faction, along with civil servants in the Treasury in a fairly closed manner, as will be discussed in the next chapter. The only policy that required votes was the A8 decision (to not place transitional controls on new accession countries). In the second reading of the European Union (accessions) Bill in 2003, with 76.6 per cent turn out from the PLP, the Party unanimously voted aye. Whether rebellions would have occurred if such changes had required legislation and thus votes is a counter-factual possibility. However, given the scarcity of rebellions on all types of policy in the first term, it seems unlikely that the reforms under study here would have caused much outcry from the PLP in any case. The closed network in which these reforms were made means that the Constituency Labour Party were very unlikely to be aware of these reforms and thus the government faced no opposition from either branches of the Party.

Whilst the reforms faced no PLP opposition when being formulated and implemented, the overarching policy shift was publicly criticised by some members of the PLP in Labour's final term of office (2005–2010) and even more so during Labour's current spell in opposition since 2010. Thus the stipulation that the Labour Party, as part of their ideological core, advocated expansive immigration should be weighed against the fact that there was not necessarily universal consensus within the Party. Rather it was the leading faction of the Party, who were proponents for economic globalisation, who steered the policy. A handful of backbenchers expressed discontent on the broad thrust of the policy reforms, claiming that the

5.7 CONSTRAINTS ON PARTY ACTION: INTERNAL DISSENT AND EXTERNAL... 139

government had not considered the local impacts of immigration in areas that received a dense concentration of this new inflow of immigrants. Nevertheless, the intra-party dissent was not fully articulated until the late 2000s, after the policy changes had transpired.

The critique of the policies from Labour MPs was conceivably motivated by electoral calculation. There was a fear that the disproportionate impact of immigration on poorer communities in Labour strongholds (such as Jon Cruddas's and Margaret Hodge's) would lead to the BNP mobilising disaffected Labour voters in the 2009 European Parliament elections (Interview with Margaret Hodge, 2012; Stevenson 2013; Ford and Goodwin 2010). In response, some Labour MPs speculated that the expansive immigration policies had not accounted for the impacts on local communities, socially, economically and culturally. Although these concerns were heightened in Labour's final term of office, the intra-party conflict on the issue of expansive immigration and its ramifications had been trailing throughout Labour's time in office:

> People like me and Jon Cruddas have constituencies which were facing massive change, we saw the political and social reactions and so we tended to take a different view. I would've said that over time, the number of people who were disconcerted by what was happening grew fairly steadily. But initially it wasn't seen as a major point of concern for the majority of people in the Party. (Interview with John Denham, 2012)

The policies had a disproportionate effect and led to concentrated immigration populations in specific areas, often-poorer areas such as Dagenham and Barking. This led to radical social and demographic changes in these communities, which, according to Hodge, Denham and Cruddas, the government did not plan or accommodate for. Fundamentally, the intra-party conflict on the issue exemplified a wider tension between national strategies on the one hand and the local impacts of policy on the other. Many Labour MPs saw this 'neglect' of local communities at the expense of expansive immigration policies as a betrayal of Labour's traditional core voters:

> These [immigration flows] are not just simply proportionally distributed across all the constituencies in Britain. There are certain ones that intensely take the strain of these pangs of extraordinary demographic changes. The poorest areas or the lowest costs housing areas and that's where we sat, we got no help whatsoever…. It [managed migration policy] doesn't have a

logic to it if you're one of those MPs in a community is just pushing through extraordinary changes which is not recognisable and you're not getting any help with. (Interview with Jon Cruddas, 2011)

Margaret Hodge similarly suggested that her constituents faced competition (or at least perceived that they were competing) with immigrants over public services and housing, which posed a contradiction in terms of Labour values, given that the Labour Party 'have a value about strong communities and strong communities means keeping people together' (Interview with Margaret Hodge, 2012).

In part the critique stemmed from a perceived inadequacy of Labour's multicultural policy. Backbenchers claimed that Labour's integration policy did not lead to dialogue between the British resident population and immigrants and that this lack of communication had created segregated communities. Moreover, some MPs suggested that the government were not investing enough in local public services in areas that had received a high concentration of immigrants:

> If you're going to be grown up enough to have this migration policy in terms of the number of people going in and out of the country, then you have to take responsibility or corresponding agenda around public services, provision, having a real time demographic audit, a real time census going on every year, housing policy, and a labour market agenda that means this isn't built on a race to the bottom because that provides really dangerous issues for community cohesion which I think we've seen. Therefore it isn't an empirical question for me of how many; it's a question of what is the broader policy mix. (Interview with Jon Cruddas, 2011)

For all three of these MPs (Cruddas, Hodge and Denham) and other Labour backbenchers, it was not so much objection to the policy of expansive immigration, but the lack of effective policy to counter local impacts of concentrated immigration, in particular costs to local public services. In addition, many MPs suggested that the social and demographic changes were too radical and too sudden for their constituents to handle.

Aside from the intra-party conflicts over the impact of these policies, there was also some disagreement within central government over how to 'handle' the issue of immigration. Blair and Brown were particularly divided as to how to manage the issue. Interviewees commented that while Blair understood the politics of asylum—indeed one interviewee claimed Blair was 'totally obsessed with asylum' (Interview with Will

Somerville, 2011)—he did not understand the politics of economic immigration:

> I think it's one of those ones where the centre or the leadership of the party was most out of touch with what was happening up and down the country... Blair in a sense understood about asylum seekers, but he didn't really understand about migration... I don't think Blair had any instinctive feeling for how migration was felt and experienced of change where the job market was... you had this disconnect with the story being told in central government and what was really happening. (Interview with Jon Denham, 2012)

A handful of interviewees alluded to a division between Blair and Brown in terms of how or if the Party could 'own' the issue. Brown, although supportive and arguably driving the earlier reforms as Chancellor, was purportedly quite fatalistic, believing Labour could never win on immigration, that immigration was a Conservative owning issue and that therefore the best approach would be to keep immigration off the political agenda:

> Brown believed very strongly that both immigration and law and order issues were issues that Labour could never win on. If you wanted to have a successful election campaign, you had to keep immigration off the agenda as far as possible. (Interview with former SpAd, 2011)

In contrast to Brown, Blair allegedly thought that if the public understood the rationale for economic immigration reform, the public would come out in favour of Labour's position. Ultimately Blair was less cynical than Brown in Labour's ability to acknowledge and own the immigration issue:

> He [Blair] was less fatalistic about the capacity of the government to manage the immigration system successfully. I think he genuinely thought that if you had the right people in the Home Office, if you had the right people managing the immigration service and agency, you had the right leader and the right policy, then actually you could control the system...but that was not a view that was universally shared across government. (Interview with former SpAd, 2011)

Central government divisions were on handling rather than the substance of policy however, and whilst there was some division within the PLP on the reforms, this was not articulated until the final term of office by which point the reforms had transpired.

5.7.2 External Constraints: 'This Will Be a Bipartisan Policy that I Hope Will Last for Many Years' (Letwin 2002)

Whilst government action is rarely defeated by its own party parliamentarians, governments seldom face such amiable conditions from opposition MPs. Parties compete to 'own' issues, and consequently opposition parties act as agents who ramp topics such as immigration up (Bale 2008) often to discredit governing parties and in a bid to win votes. Until the formation of the Conservative led Coalition government in 2010, Britain had long been regarded as a two-party system. The analysis here therefore concentrates on the government's chief and official (shadow cabinet) opposition, the Conservative Party.

Traditionally right wing parties dominate and 'own' the issue of immigration, at least in terms of competency and border control (Green and Hobolt 2008). Yet as discussed in Chapter 2, immigration is a problematic issue for centre-right parties, such as the British Conservative Party, as it exposes the tensions between the 'identity right' and the 'business right' (Bale 2008). Nevertheless, the Conservative Party pursued restrictive immigration policies throughout the post-war period. In contrast to this decidedly restrictive stance, the Conservative opposition scarcely objected to Labour's expansive policy reforms in the late 1990s and early 2000s and were in fact largely supportive.

This submission can partly be attributed to the Party's generally 'futile period in opposition' between 1997 and 2001 (Collings and Seldon 2001, 624). Whilst the 1997 defeat was 'not unexpected, its scale was devastating' (Robinson 2012, 89). The Conservative Party had lost its long-cherished reputation of economic competence following the Exchange Rate Mechanism crisis in 1992, which 'almost certainly undermined any subsequent economic policy initiatives and contributed substantially to Labour's [1997] electoral landslide' (Farrell and Webb 2002, 23). Indeed, the European issue proved to be a 'profoundly divisive issue for the ruling Conservative Party' (Taggart 1998, 365). Numerous Tory MP scandals in the 1990s saw the Party labelled 'the Party of sleaze' (Farrell et al. 1998). Coupled with the deep and publicly visible divisions over Europe, the Party had lost the public's faith by the 1997 General Election (Farrell et al. 1998; McAllister and Studlar 2000). The heavy electoral defeat left the Party intensely divided and lacking in policy direction and clarity on all fronts (Collings and Seldon 2001, 628).

The Labour Party on the other hand held the largest majority in the House of Commons in the post-war period, making the government especially powerful and able to govern (and pass policies) with supreme autonomy. Blair has been compared with two agenda-setting prime ministers of the past century, Attlee and Thatcher, and his party governed under similar conditions, which contributed to Labour's dominance, including: a long period of office, a large parliamentary majority, weak opposition and a favourable climate of opinion (Kavanagh 2005, 4). Alongside a booming economy 'it would be hard to imagine more propitious circumstances than those in which Tony Blair's New Labour government swept to power in May 1997' (Brown 2011, 403). In contrast, the Conservative opposition was divided, uncoordinated and fairly debilitated following such a heavy defeat.

The generally weak position of the Conservatives aside, the parliamentary debates on immigration in the late 1990s and early 2000s (during which the economic immigration reforms were established) were dominated by asylum in any case. Therefore any opposition from the Conservatives was focused on changes to the asylum policy, as Blunkett's former SpAd surmised: 'I mean the opposition weren't particularly focused on this [economic immigration], no one was!' (Interview with former SpAd, 2012). Former IPPR associate Danny Sriskandarajah similarly commented that the lack of attention, both politically and publicly, on economic immigration partly explains how the Labour government were able to pass such radical policies:

> Because the focus in the early 2000s was so much on asylum, they could get away with it. I mean nobody was looking at net migration in the way they do now, as a sort of overall number based target. People here were just thinking of the number of asylum seekers, what was happening in labour market terms wasn't that interesting.

In turn, press attention was focused on asylum seekers and refugees in the late 1990s and early 2000s, and until the 2004 A8 decision, economic immigration 'largely slipp[ed] under the radar' (Mulvey 2011, 1485). The right-wing press were particularly attentive (and hostile) to asylum seekers, with the *Daily Express* and the *Daily Mail* covering more stories on asylum than anything else in 2002 (Greenslade 2005, 21). Nonetheless, in terms of economic policy, the press were inattentive:

There was no great anxiety from the press. At the very least there was no underlying hostility in the press except on particular things like asylum. If anything there was a positive view, but that did change by 2005. (Interview with David Goodhart, 2012)

The lack of press attention and political opposition on the government's economic immigration reforms gave the Party sufficient autonomy to pass these policies with relative ease. We now turn to the Conservative's lack of opposition to Labour's reforms by chronologically tracing the Party's framing of immigration and reaction to Labour's reforms whilst in opposition, with a specific focus on the period 1997–2005 when the reforms under study here were made.

5.7.2.1 1997–2005: Asylum Crisis
In the run-up to the 1997 General Election immigration barely warranted a mention in the Conservative Party's manifesto. The only suggestion was to maintain the post-war bipartisan consensus that 'firm but fair immigration controls underpin good race relations' (Conservative Party 1997). Beyond these broad statements that contained little policy substance, immigration and asylum were not on the political programme. This is hardly surprising when contrasted with the likewise scarcity of immigration content in Labour's manifesto. Immigration was simply not on the political agenda in 1997 (Kriesi and Frey 2008; Saggar 1997a, b).

Within the House of Commons, the Conservative's opposition to immigration reforms was weak and uncoordinated. The opposition focused their attention on the asylum backlog as well as so-called 'bogus asylum seekers', strengthening border controls and, following the Amsterdam Treaty, whether the UK would participate in the Schengen acquis. Following the 1998 White Paper *Fairer, Faster, Firmer* the Conservatives were broadly supportive of Home Secretary Jack Straw's new measures on asylum (Hansard 27 July 1998). As Straw reflected: 'No one on the Opposition Benches condones bogus applications for asylum: everyone condemns them' (Hansard 27 July 1998, col 40). Thus unsurprisingly the measures set out in the White Paper to curb abuses in the asylum system were met with cross-party support.

In the run-up to the 2001 election, asylum began to creep further up the political agenda. Immigration and asylum were one of the few issues that the Conservatives enjoyed a lead on, and the Party unsurprisingly chose to exploit this in the 2001 campaign (Bale 2010, 122). Leader of the opposition William Hague adopted a tough rhetoric on asylum, claiming that 'People are arriving in Britain armed with expert knowledge of how to exploit our asylum laws; what to say on arrival; how to string out appeals and how to remain here if their cases are eventually turned down' (Hague cited in BBC 2000). In contrast to the Conservative's 1997 manifesto, the Party's 2001 manifesto contained far more policy substance on asylum, suggesting that 'Britain has gained a reputation as a soft touch for bogus asylum seekers' (Conservative Party 2001). However, while the Conservatives addressed asylum, there was no mention of economic immigration.

Whilst the Party initially took advantage of their lead on immigration, as the campaign progressed the Party were accused by the *Sun* newspaper of 'flirting with extremism' (Sun 2001). On pain of being accused of prejudice again, Hague signed a Commission for Racial Equality (CRE) pledge committing politicians not to 'play the race card' (Guardian 2001). Whilst the campaign served to reinforce the Party's long held lead on immigration, the Party could not lift it above those issues 'which counted most and on which Labour enjoyed massive leads' (Bale 2010 130). Therefore, whilst 'Hague ran quite a traditional tough immigration policy in the 2001 election and of course didn't get very far, in a sense that's an indication that it wasn't hugely salient at that point' (Interview with former SpAd, 2011). Although the tough rhetoric on immigration and asylum possibly won the Party a few votes, the campaign may have done the Party more harm than good, as 'it provided its opponents with yet another chance to brand it as extreme, obsessive, and old fashioned' (Bale 2010, 130).

Throughout the late 1990s and early 2000s, any debate on immigration in the House was consistently deflected by discussions on the handling of asylum, epitomised by MP Julian Brazier's question to the Home Secretary within minutes of a debate on immigration beginning:

> Everybody accepts the terrific benefits that this country has enjoyed over the years and centuries from the immigration of people with key skills, but is not this scheme one more example of the Government trying to paper over the disaster that is their asylum policy? (Julian Brazier, Hansard 28 October 2002)

While the saliency of asylum may have been rising, objections to the economic immigration policies were scarce. When Home Secretary David

Blunkett announced proposals to expand economic immigration routes—by liberalising the work permit criteria, developing 'fast track' entry routes for those with high skills and allowing students to remain in the UK for work post-study—his announcement was met with limited opposition. To the extent that there were debates on the economic immigration reforms, the Conservatives appeared to be broadly supportive of the new schemes:

> Neither the Home Secretary's Labour predecessor nor his Conservative predecessors moved as well in this direction as he has sought to do. This will be a bipartisan policy that I hope will last for many years. (Oliver Letwin, Hansard 24 April 2002)

The third largest party, the Liberal Democrats, were likewise supportive, concurring that 'migration is net financially beneficial' (Simon Hughes, Hansard 19 March 2003) with current deputy leader of the Party Simon Hughes being particularly enthusiastic:

> The Home Secretary sensibly responded to the call for a more inclusive and widespread review of immigration policy so that more people can enter the country lawfully to do the jobs that British society and our economy need them to do. (Hansard 24 April 2002)

A few Conservative MPs did however raise concerns about the potential number of immigrants that the new managed migration scheme could potentially bring to Britain, including former Prime Minister Cameron (David Cameron, Hansard 19 March 2003; Peter Lilley, Hansard 24 April 2002). The Conservatives appeared to accept and acknowledge the government line that immigration was an economic good, but argued that the 'economic benefits of migration may be outweighed by its negative social impact' (Boswell 2009, 120; Hansard 19 March 2003). It should however be noted that there was a lack of parliamentary debate for much of the reforms in the first instance because they did not require primary legislation.

Although opposition from the Conservative Party to migration schemes, such as the expansion of work permits, was limited, the A8 decision played out slightly differently. The Conservatives could have mobilised anti-immigrant public preferences, and even Euroscepticsm, by framing the A8 decision as an expansive and uncontrollable immigration policy. Yet the decision was mainly debated in foreign policy terms and the

5.7 CONSTRAINTS ON PARTY ACTION: INTERNAL DISSENT AND EXTERNAL... 147

Conservatives were contrarily supportive, demonstrated by 126 Conservative MPs voting aye and only two members rebelling (with a 78.5 per cent turnout) in the second reading of the Bill in 2002. Conservative MPs made frequent reference to the fact that, while in government, they had pushed for Central and Eastern Europe (CEE) pre-accession negotiations. Opposition Spokesman for Foreign Affairs Richard Spring articulated this as part of his opening statement on the first debate of EU enlargement:

> I can only repeat that, when in government, our party was one of the first advocates of enlargement to embrace Central and Eastern Europe...We warmly welcome the participation of these countries in the European Union, and, of course that implies freedom of movement. (Hansard 5 June 2003, cols 346–75)

He originally began the debate by calling for transitional controls for the first year of accession, claiming that immigration flows from the A8 would be far higher than predicted. Yet by the end of the debate Spring withdrew his objection [and amendment 4 (transitional controls)]:

> The Minister has said that safeguards are in place and she has spelled them out. I accept and I am grateful to her for that...I endorse entirely her point that we want to send out a positive message to the accession countries. We were entitled to ask for clarification, but the overriding importance of the Bill was shown on Second Reading, when something historically extraordinary happened in this Chamber: unanimous acceptance of the Bill. In that context, we will not seek to divide the Committee. The spirit of our support across the party political divide for the accession countries is so deeply rooted in the thinking of successive Governments that we do not propose to take these amendments any further. (Hansard 5 June 2003, vol 406)

This statement reflects the bipartisan support of the A8 decision. Britain was a keen supporter and indeed a 'driver' of CEE accession (Schimmelfennig 2001). The interest in CEE accession was principally due to trading ties and a foreign policy interest in forging alliances with CEE states at the EU level, which under the early 1990s Conservative governments 'appears to have been based on the calculation that an extensive "widening" of the Community would prevent its further "deepening" and might even dilute the achieved level of integration' (Schimmelfennig 2001, 71). As a result, Thatcher was one of the biggest advocates of CEE accession: 'We can't say in one breath that they [CEE states] are part of Europe and in the next our European club is so exclusive that we won't admit them' (Thatcher 1990, 3). As a consequence of Britain's foreign policy interests, all parties were supportive of unfettered labour market access to new CEE citizens, almost as a symbolic gesture to Britain's expectant allies. Arguably if the Tories had won office in 2001 they would have made the same decision (Persin 2007, 21).

However, as 1 May 2004 approached, the Tories' enthusiasm for CEE accession seemed to wane and some, albeit limited, opposition was raised as to how many A8 citizens would come to Britain (Ann Winterton, Hansard, 5 June 2003). Objections to EU enlargement were mainly concerned with 'welfare shopping', the impact on agricultural communities along with fears of a possible influx of 'Roma' people (although the latter concern was initially raised by Labour MP Denzil Davies). The principal contention was that A8 migrants would come to the UK to acquire social benefits and that this would place great pressure on housing and public services. The solution proposed by Conservative MPs, such as David Davis, was to use the work permit system to regulate A8 migration (Hansard 23 February 2004, vol 418). Therefore the objection from the Conservatives was the method of monitoring immigration from the A8 states rather than the decision itself. In response, Blair quickly announced that citizens of the EU's new member states would have their access to UK welfare payments severely restricted (Bale 2010, 207), and as a result of the Tories' politicisation of potential welfare shopping (combined with media pressure) the government was forced to rapidly develop and implement the Workers Registration Scheme (WRS).

With a new party leader at the end of 2003 came a new party strategy. In contrast to his predecessor Iain Duncan-Smith, newly appointed leader of the Conservative Party Michael Howard called for transitional controls on the A8 countries. Howard outlined his objections in his 'British Dream' speech in Burnley (scene three years previously of race riots) in February 2004, a mere 3 months before accession, which suggests that these objections were likely to be politically calculated, partly because the BNP had gained some success in recent local elections (Gajewska 2006, 391). However, business groups such as the BCC were very critical of this protectionist line (Bale 2010, 207), and in turn Howard soon went quiet on imposing transitional controls.

5.7.2.2 2005 Break Point: It's Not Racist to Impose Limits on Immigration

The 2005 election campaign saw a harder, tougher and arguably more populist and orthodox stance from the opposition on immigration and asylum. The Party, or more specifically leader Howard and shadow immigration minister David Davis, claimed in interviews and a full page ad in the *Sunday Telegraph* that immigration was 'unlimited' under Labour, which was allowing 'a city the size of Peterborough' to settle in the UK each year (Bale 2010, 242). The Tories also employed a measure of 'dog-whistle politics' (Geddes and Tonge 2005) epitomised in their 2005 manifesto title, *Are you thinking what we're thinking?*, where leader Michael Howard condemned the 'out-of-control immigration system' (Conservative Party 2005). In contrast to the mere paragraph on immigration in the 2001 election manifesto, the Conservative Party's 2005 manifesto dedicated a whole chapter to the Party's plans on immigration with the infamous title 'It's not racist to impose limits on immigration'.

For the first time in opposition, the Conservatives proposed an explicit limit on immigration in their manifesto, arguing that 'refusing to set a limit on new migrants is irresponsible politics' (Conservative Party 2005). The manifesto proposed: withdrawing from the 1951 Geneva Convention; 24-hour surveillance at ports and the re-establishment of embarkation controls; a points-based system for work permits; and an annual immigration limit to include asylum seekers (Conservative Party 2005). The Tories also insisted that a Conservative government would make all immigrants coming to Britain for more than a year withstand screening for HIV/AIDS, hepatitis and TB to protect public health. This hard stance went down well with the press and the public, with 97 per cent of *Sun* readers for example backing Howard's proposed curbs on migrants (Bale 2010, 244).

Blair responded to Howard's calls for curbing immigration with 'customary skill', assuring Howard that 'I am not accusing you of being a racist… you are just a shameless opportunist' (Blair quoted in BBC 2005a). Blair's rebuttal in a speech in Dover (Blair 2005) likewise served as a successful counter-attack on the Tories criticisms of Labour's policies, and Blair 'effectively shut down the Tory attack, because our position was sophisticated enough—a sort of "confess and avoid", as lawyers' say—we won out' (Blair 2010, 524). The polls immediately following this attack implied it was a line that resonated with the public, with only 36 per cent of respondents agreeing that Howard 'genuinely believes immigration should be limited', and 58 per cent believing Howard focused on the issue of immigration because 'he desperately wants to win votes for his party' (Bale 2010, 243).

While the Conservatives were evidently not successful in their campaign, it is clear that by 2005 the bipartisan consensus of defusing and depoliticising immigration had ended, and immigration was now very much an acceptable issue for the political battle. Having said this, the Conservative's lack of electoral success at both the 2001 and 2005 elections perhaps indicates that although immigration was rising up the political agenda, it was not yet an election defining issue: 'Labour won those elections in a big way, despite the Conservatives playing the migration card in both elections, it might've led them [the Labour Party] to think well it plays big but it's not that big' (Interview with Tim Finch, 2011). Nonetheless, the Conservatives' attempt to politicise immigration did have an impact on the Labour Party, to the extent that Blair was purportedly worried that Labour would lose the 2001 election on asylum (Interview with Tim Finch, 2011; Interview with Will Somerville, 2011; Seldon et al. 2007).

With the Conservatives taking an even tougher stance on immigration and asylum in the 2005 General Election, Labour was now preoccupied with how to handle the issue (Interview with former SpAd, 2012). Labour's five-year plan on immigration, which revealed the proposal for PBS, was therefore intentionally published a few months before the 2005 General Election to reassure the public on immigration and asylum. The Strategy Paper thus coincided with Labour's electoral strategy for the 2005 Election: 'We sought to neutralise these issues by dealing with them systematically through the five year plans and counterattacking the Tories' (Gould 2007, 21).

It is fair to say that the opposition's tougher position and rhetoric on immigration had some impact on Labour's approach; that the five-year strategy published months before the election contained proposals for tougher measures such as biometric ID cards and a reduction of protection granted to refugees (Home Office 2005) is no coincidence. As discussed in Chapter 3, Labour ran a series of focus groups during the 2005 campaign, led by strategists and polling advisors Philip Gould and Deborah Mattinson. The polling found that anti-immigrant public sentiment was widespread, and as a result the government began 'rowing back' on immigration.

Conservative opposition on immigration was nonetheless weak between 1997 and 2005, when the policy reforms transpired, and attempts to take a harder stance on immigration and asylum, for the most part, backfired politically and electorally. All parties criticised Howard's immigration plans. Even the leader of the UK Independence Party (UKIP) at the time, Roger Knapman, commented that Tory immigration policy was 'so unworkable it was almost laughable' (Knapman quoted in BBC 2005b). The problem for the Conservatives was the government's tough line on asylum. Labour's tougher position on immigration, if only in rhetoric, effectively hijacked the centre-right political space. By co-opting opposition proposals 'Blair pushed the Conservatives onto extremist grounds and forced them to present unfeasible proposals' (Carvalho 2012, 161). This tough rhetoric from the government meant that a hard-line was strategically questionable for the opposition:

> The Party's big lead on asylum and immigration did not, modernizers pointed out, translate into boosting its overall support because it was mostly preaching to the converted and to less educated, less well-off people who would, if they bothered to vote at all, nonetheless continue to vote Labour while it delivered the goods on welfare and the economy. (Bale 2010, 171)

The Party were seeking to modernise and disassociate themselves from the 'nasty party' image they had acquired previously with 'Hagueite headline chasing' (Bale 2010, 171). Yet conversely Howard's campaign only served to reinforce such impressions.

The lack of an electorally significant far-right party also allowed the Labour governments to pursue expansive immigration policies. Whilst much of the rest of Western Europe were seeing electorally successful far-right parties in the 2000s (such as in The Netherlands, Switzerland and France), Britain's lack of an electorally successful populist far-right party meant that a forceful anti-immigrant rhetoric was never legitimised, at least not on ethno-cultural lines. Where far right parties are electorally successful, centre parties of the left and right often co-opt proposals and reinforce a tough rhetoric on immigration (Schain 2008; Bale et al. 2010). While the BNP saw some relative success at the 2009 European election (gaining two MEPs) and local elections, this success was years after the policy changes had transpired and was, after all, in secondary elections. Neither UKIP nor the BNP managed to gain a seat in the House of Commons between 1997 and 2010. The politics of immigration look very different in countries where the far-right have gained electoral success, and the lack of credible far-right party during the Labour years meant that the immigration debate was perhaps less toxic than it may have otherwise been.

The ineffectual opposition from the established parties (principally the Conservative Party) coupled with an absent far-right party contributed to a conducive political environment for passing expansive policies. Likewise, a consensual non-rebelling PLP meant that the core leadership faced limited constraints and could operate fairly autonomously. However, weak opposition from the established parties cannot explain *why* the Labour governments ratified these policies in the first instance. It rather explains *how* such radical reforms transpired without political contention.

5.8 Conclusion

The *Making of a Migration State* sets out to explain why the policy change occurred and which actors were steering the policy reforms. In this chapter we have considered whether the ideology of the governing party, and the party political context in which these policies were made, could explain the policy shift. We have seen that a change to a centre-left governing party was a necessary condition for policy change. If we consider the counterfactual possibility of the Conservative Party remaining in power in 1997, it is highly unlikely that the policy changes would have occurred. However, centre-left parties do not unequivocally favour expansive immigration policies; comparisons across Western Europe, as well as Labour's past (and current policy), demonstrate that this is not the case. Ideologically there is no fixed or given position on immigration for the centre left. Nonetheless, being part of a centre-left party (in contrast to a centre-right party) does seem to shape political elites' preferences on immigration to some extent (Lahav 1997) and can therefore shape public policy. While a change to a centre-left government was a necessary condition for an expansive policy, it cannot be said to be a sufficient explanation.

An analysis of the ideology of the governing party goes some way to explaining the policy shift, and in this sense the 'politics-matter' school of thought (Imbeau et al. 2001, 1) has resonance in explaining this case of policy change. However, given that the policy outcomes were discordant with the general publics' preferences (Ipsos Mori 2004; Saggar 2003), this was clearly not a policy made to win votes. It is rather the party ideology, as opposed to electoral strategies, which provides an explanation for the policy shift. The ideological re-foundation of the Labour Party proved to have a powerful effect on the preferences of certain leading Labour ministers. A major component of the Party's modernisation was the acceptance and endorsement of globalisation. The rebranded Labour Party claimed that globalisation was both inevitable and an intrinsically positive thing. Immigration was no longer seen as a threat to national workers' wages. Conversely, immigration was seen as part and parcel of this new globalised economy and a tool to aid labour market flexibility. These ideas permeated through the leading factions of the Labour Party and fundamentally changed their immigration policy preferences. That globalisation, which 'was used ideologically as the raison d'être of New Labour' (Cole 1998, 323), proved to be a critical factor for the policy change, albeit in an intangible way.

Labour's ideological overhaul was also important because of the way in which the ideology was mediated by elites in a convoluted way through institutional reform and departmental agendas. There was no conscious strategy to expand immigration, but the logic behind Labour's economic policy, modernisation agenda and ideal of the competition state was manifested in both policymaking practices and departmental agendas across Whitehall. As we shall see in the next chapter, it is here between the nexus of party ideology and its expressions.

REFERENCES

Balch, A. (2010). *Managing Labour Migration in Europe: Ideas, Knowledge and Policy Change*. Manchester: Manchester University Press.

Bale, T. (1999). *Sacred Cows and Common Sense: The Symbolic Statecraft and Political Culture of the British Labour Party*. Surrey: Ashgate Publishing.

Bale, T. (2008). Turning Round the Telescope. Centre-Right Parties and Immigration and Integration Policy in Europe. *Journal of European Public Policy, 15*(3), 315–330.

Bale, T. (2010). *The Conservative Party: From Thatcher to Cameron*. Cambridge: Polity Press.

Bale, T., Green-Pedersen, C., Krouwel, A., Luther, K. R., & Sitter, N. (2010). If You Can't Beat Them, Join Them? Explaining Social Democratic Responses to the Challenge from the Populist Radical Right in Western Europe. *Political Studies, 58*(3), 410–426.

Balls, E. (1998). Open Macroeconomics in an Open Economy: Scottish Economic Society/Royal Bank of Scotland Annual Lecture, 1997. *Scottish Journal of Political Economy, 45*(2), 113–132.

BBC. (2000, April 30). Hague Fuels Asylum Row. *BBC*. Available from: http://news.bbc.co.uk/1/hi/uk_politics/731696.stm. Accessed 19 July 2013.
BBC. (2005a, January 26). Howard and Blair Clash on Asylum. *BBC*. Available from: http://news.bbc.co.uk/1/hi/uk_politics/4209001.stm. Accessed 11 Mar 2013.
BBC. (2005b, April 10). Tory Leader Attacks Asylum Seekers. *BBC*. Available from: http://news.bbc.co.uk/1/hi/uk_politics/vote_2005/frontpage/4428517.stm. Accessed 11 Mar 2013.
Bell, D. (1962). *The End of Ideology: On the Exhaustion of Political Ideas in the Fifties: With "The Resumption of History in the New Century"*. Harvard: Harvard University Press.
Blair, T. (1996). *New Britain: My Vision of a Young Country*. London: 4th Estate.
Blair, T. (2000a, January 18). *Speech at the World Economic Forum Meeting*. Davos, Switzerland.
Blair, T. (2000b, March 28). *Tony Blair's Britain Speech*. Available from: http://www.guardian.co.uk/uk/2000/mar/28/britishidentity.tonyblair. Accessed 19 Dec 2011.
Blair, T. (2004, April 27). *Speech to Confederation of British Industry on Migration*, London. Available from: http://www.theguardian.com/politics/2004/apr/27/immigrationpolicy.speeches. Accessed 21 Mar 2010.
Blair, T. (2005, April 22). *Speech on Asylum and Immigration*. Dover. Available from: http://www.theguardian.com/politics/2005/apr/22/election2005.immigrationandpublicservices. Accessed 5 June 2011.
Blair, T. (2006, September 12). *Prime Minister's Address to the Trade Union Congress*. Available from: http://www.tuc.org.uk/about-tuc/congress/congress-2006/prime-ministers-address-congress-2006. Accessed 16 Mar 2010.
Blair, T. (2010). *Tony Blair: A Journey*. London: Hutchinson.
Blunkett, D. (2002). Integration with Diversity: Globalisation and the Renewal of Democracy and Civil Society. Essay collection on *Rethinking Britishness*. London: Foreign Policy Centre. Available from: http://fpc.org.uk/articles/182. Accessed 5 July 2010.
Blunkett, D. (2003). *Towards a Civil Society*. London: IPPR.
Boswell, C. (2009). Knowledge, Legitimation and the Politics of Risk: The Function of Research in Public Debates on Migration. *Political Studies, 57*(1), 165–186.
Brown, G. (2001). The Conditions for High and Stable Growth and Employment. *The Economic Journal, 111*(471), 30–44.
Brown, G. (2007, July 3). Oral Answers to Questions—Foreign and Commonwealth Affairs: Constitutional Reform. *Hansard House of Commons Questions*.
Brown, W. (2011). Industrial Relations in Britain Under New Labour, 1997–2010: A Post-mortem. *Journal of Industrial Relations, 53*(3), 402–413.
Bulpitt, J. (1996). Historical Politics: Leaders, Statecraft and Regime in Britain at the Accession of Elizabeth II. *Contemporary Political Studies, 6*(2), 1093–1106.

Burnham, P. (2001). New Labour and the Politics of Depoliticisation. *The British Journal of Politics and International Relations,* 3(2), 127–149.
Carvalho, J. (2012). *Bringing Politics Back in: the Impact of Extreme-Right Parties on Immigration Policy in the UK, France and Italy During the 2000s* (PhD thesis). University of Sheffield.
Cerny, P. G., & Evans, M. (2004). Globalisation and Public Policy Under New Labour. *Policy Studies,* 25(1), 51–65.
Cole, M. (1998). Globalisation, Modernisation and Competitiveness: A Critique of the New Labour Project in Education. *International Studies in Sociology of Education,* 8(3), 315–333.
Collings, D., & Seldon, A. (2001). Conservatives in Opposition. *Parliamentary Affairs,* 54(4), 624–637.
Conservative Party. (1997). *1997 Conservative Party General Election Manifesto: You Can Only Be Sure with the Conservatives.* London: Conservative Party.
Conservative Party. (2001). *2001 Conservative Party General Election Manifesto: Time for Common Sense.* London: Conservative Party.
Conservative Party. (2005). *Conservative Party General Election Manifesto: Are You Thinking What We're Thinking? It's Time for Action.* London: Conservative Party.
Consterdine, E. & Hampshire, J. (2014). Immigration Policy Under New Labour: Exploring a Critical Juncture. *British Politics,* 9(3), 275–296.
Cowley, P. (2002). *Revolts and Rebellions: Parliamentary Voting Under Blair.* London: Politico's.
Cowley, P. (2005). *The Rebels: How Blair Mislaid His Majority.* London: Politico's.
Cowley, P., & Stuart, M. (2003). In Place of Strife? The PLP in Government, 1997–2001. *Political Studies,* 51(2), 315–331.
Daniels, P. (2003). From Hostility to "Constructive Engagement": The Europeanisation of the Labour Party. In A. Chadwick & R. Heffernan (Eds.), *The New Labour Reader* (pp. 225–231). Cambridge: Polity Press.
Driver, S. (2011). *Understanding British Party Politics.* Cambridge: Polity Press.
Driver, S., & Martell, L. (2002). *Blair's Britain.* Cambridge: Polity Press.
Dyson, S. B. (2006). Personality and Foreign Policy: Tony Blair's Iraq Decision. *Foreign Policy Analysis,* 2, 289–306.
Farrell, D. M., & Webb, P. (2002). Political Parties as Campaign Organizations. In R. Dalton & M. Wattenbrug (Eds.), *Parties Without Partisans: Political Change in Advanced Industrial Democracies'* (pp. 102–128). Oxford: Oxford University Press.
Farrell, D. M., McAllister, I., & Studlar, D. T. (1998). Sex, Money and Politics: Sleaze and the Conservative Party in the 1997 Election. *British Elections and Parties Review,* 8(1), 80–94.
Fielding, S., & Geddes, A. (1998). The British Labour Party and "Ethnic Entryism": Participation, Integration and the Party Context. *Journal of Ethnic and Migration Studies,* 24(1), 57–72.

Ford, G. (1991). *Labour Party Archive, NEC Annual Conference Report 1990*, 30/09/90–05/10/90. Labour History Archive Study Centre [LHASC].
Ford, R., & Goodwin, M. (2010). Angry White Men: Individual and Contextual Predictors of Support for the British National Party. *Political Studies, 58*(1), 1–25.
Freeden, M. (2003). The Ideology of New Labour' Democracy. In A. Chadwick & R. Heffernan (Eds.), *The New Labour Reader* (pp. 43–49). Cambridge: Polity Press.
Gajewska, K. (2006). Restrictions in Labour Free Movement After the EU Enlargement 2004: Explaining Variation Among Countries in the Context of Elites' Strategies Towards the Radical Right. *Comparative European Politics, 4*(4), 379–398.
Geddes, A., & Tonge, J. (2005). *Decides: The 2005 General Election*. London: Palgrave Macmillan.
Giddens, A. (2003). The Third Way: The Renewal of Social Democracy. In A. Chadwick & R. Heffernan (Eds.), *The New Labour Reader* (pp. 34–39). Cambridge: Polity Press.
Giddens, A. (2007, June 28). It's Time to Give the Third Way a Second Chance. *The Independent*. Available from: http://www.politicsforum.org/forum/viewtopic.php?f=28&t=79097. Accessed 21 Apr 2011.
Givens, T., & Luedtke, A. (2005). European Immigration Policies in Comparative Perspective: Issue Salience, Partisanship and Immigrant Rights. *Comparative European Politics, 3*(1), 1–22.
Glyn, A., & Wood, S. (2001). Economic Policy Under New Labour: How Social Democratic Is the Blair Government? *Political Quarterly, 72*(1), 50–66.
Gould, P. (2007). Labour's Political Strategy. In D. Wring, J. Green, R. Mortimore, & S. Atkinson (Eds.), *Political Communications: The General Election Campaign of 2005* (pp. 17–25). Basingstoke: Palgrave Macmillan.
Green, J., & Hobolt, S. (2008). Owning the Issue Agenda: Party Strategies and Vote Choices in British Elections. *Electoral Studies, 27*(3), 460–476.
Greenslade, R. (2005). *Seeking Scapegoats: The Coverage of Asylum in the UK Press*. London: IPPR.
Hall, S. (2003). The Great Moving Nowhere Show. In A. Chadwick & R. Heffernan (Eds.), *The New Labour Reader* (pp. 82–88). Cambridge: Polity Press.
Hampshire, J. (2006). Immigration and Race Relations. In P. Dorey (Ed.), *The Labour Governments, 1964–70* (pp. 309–330). Oxford: Routledge.
Hampshire, J., & Bale, T. (2015). New Administration, New Immigration Regime: Do Parties Matter After All? A UK Case Study. *West European Politics, 38*(1), 145–166.
Hansard HC Deb. vol. 317 cols. 25–54, 27 July 1998.
Hansard HC Deb. vol. 384 cols. 341–4224, 24 April 2002.
Hansard HC Deb. vol. 391 cols. 534–6, 28 October 2002.

Hansard HC Deb. vol. 401 cols. 241–263, 19 March 2003.
Hansard HC Deb. vol. 406 cols. 346–75, 5 June 2003.
Hansard HC Deb. vol. 418 cols. 23–36, 23 February 2004.
Hansen, R., & King, D. (2000). Illiberalism and the New Politics of Asylum: Liberalism's Dark Side. *Political Quarterly, 71*(4), 396–403.
Harrison, S. (2002). New Labour, Modernisation and the Medical Labour Process. *Journal of Social Policy, 31*(3), 465–485.
Heffernan, R. (2000). *New Labour and Thatcherism: Political Change in Britain*. Hampshire: Macmillan Press.
Hennessy, P. (2005). Rulers and Servants of the State: The Blair Style of Government 1997–2004. *Parliamentary Affairs, 58*(1), 6–16.
Home Office. (2002). *Secure Borders, Safe Havens*. London: Home Office.
Home Office. (2005). *Controlling Our Borders: Making Migration Work for Britain—Five Year Strategy for Asylum and Immigration*. London: Home Office.
Imbeau, L. M., Pétry, F., & Lamari, M. (2001). Left–Right Party Ideology and Government Policies: A Meta-Analysis. *European Journal of Political Research, 40*(1), 1–29.
Ipsos-Mori. (2004). Can We Have Diversity and Trust? Available from: http://www.ipsos-mori.com/researchpublications/researcharchive/793/Can-We-Have-Trust-And-Diversity.aspx. Accessed 14 Feb 2013.
Jordon, B. (1991). *Labour Party Archive, NEC Annual Conference Report 1991*, 29/09/91–5/10/91. Labour History Archive Study Centre [LHASC].
Kavanagh, D. (2005). The Blair Premiership. In A. Seldon & D. Kavangh (Eds.), *The Blair Effect 2001–05* (pp. 3–20). Cambridge: Cambridge University Press.
Kinnock, N. (1991). *Labour Party Archive, NEC Annual Conference Report 1991*, 29/09/91–5/10/91. Labour History Archive Study Centre [LHASC].
Koopmans, R., Michalowski, I., & Waibel, S. (2012). Citizenship Rights for Immigrants: National Political Processes and Cross-National Convergence in Western Europe, 1980–2008. *American Journal of Sociology, 117*(4), 1202–1245.
Kriesi, H., & Frey, T. (2008). The United Kingdom: Moving Parties in a Stable Configuration. In H. Kriesi, E. Grande, R. Lachat, M. Dolerzal, S. Bornschier, & T. Frey (Eds.), *West European Politics in the Age of Globalization* (pp. 183–208). Cambridge: Cambridge University Press.
Kriesi, H., Grande, E., Lachat, R., Dolezal, M., Bornschier, S., & Frey, T. (2008). Globalization and Its Impact on National Spaces of Competition. In H. Kriesi, E. Grande, R. Lachat, M. Dolerzal, S. Bornschier, & T. Frey (Eds.), *West European Politics in the Age of Globalization* (pp. 1–23). Cambridge: Cambridge University Press.
Labour Party. (1997). *Labour into Power: A Framework for Partnership; New Labour: Because Britain Deserves Better*. London: Labour Party.

Labour Party. (2001). *2001 Labour Party General Election Manifesto: Ambitions for Britain*. London: Labour Party.
Labour Party. (2005). *Labour Party General Election Manifesto: Britain Forward Not Back*. London: Labour Party.
Lahav, G. (1997). Ideological and Party Constraints on Immigration Attitudes in Europe. *Journal of Common Market Studies, 35*(3), 377–406.
Leggett, W. (2000). New Labour's Third Way: From "New Times" to "No Choice". *Studies in Social and Political Thought, 3*, 19–31.
Letwin, O. (2002, April 24). *Hansard HC Deb* (Vol. 384, Cols. 341-4224).
LHASC/LRPD. (1994). Domestic and International Policy Committee 'Resolutions from Affiliated Organisations: November/December 1993, PD: 3419/January 1994
LHASC/NEC Annual Conference Report. (1990).
LHASC/NEC Annual Conference Report. (1991).
Lynch, P. (2000). The Conservative Party and Nationhood. *Political Quarterly, 71*(1), 59–67.
Mandelson, P. (2010). *The Third Man*. London: Harper Press.
Martell, L., & Driver, S. (2006). *New Labour*. Cambridge: Polity Press.
McAllister, I., & Studlar, D. T. (2000). Conservative Euroscepticism and the Referendum Party in the 1997 British General Election. *Party Politics, 6*(3), 359–371.
McLean, I. (2001). *Rational Choice and British Politics*. Oxford: Oxford University Press.
Menz, G. (2008). *The Political Economy of Managed Migration*. Oxford: Oxford University Press.
Mulvey, G. (2011). Immigration Under New Labour: Policy and Effects. *Journal of Ethnic and Migration Studies, 37*(9), 1477–1493.
OECD. (2009). *International Migration: The Human Face of Globalisation*. Paris: OECD.
Oppermann, K. (2008). The Blair Government and Europe: The Policy of Containing the Salience of European Integration. *British Politics, 3*(2), 156–182.
Persin, D. (2007). Free Movement of Labour: UK Responses to the Eastern Enlargement and GATS Mode 4. *Journal of Word Trade, 42*(5), 1–54.
Richards, D., & Smith, M. (2010). Back to the Future: New Labour, Sovereignty and the Plurality of the Party's Ideological Tradition. *British Politics, 5*(3), 239–264.
Robinson, E. (2012). *History, Heritage and Tradition in Contemporary British Politics: Past Politics and Present Histories*. Manchester: Manchester University Press.
Roche, B. (2000, September 11). *UK Migration in a Global Economy*. Speech to IPPR.

Saggar, S. (1997a). The Dog that Didn't Bark: Immigration, Race and the Election. In A. Geddes & J. Tonge (Eds.), *Labour's Landslide: The 1997 General Election* (pp. 147–163). Manchester: Manchester University Press.
Saggar, S. (1997b). Racial Politics. In P. Norris & N. T. Gavin (Eds.), *Britain Votes 1997* (pp. 185–199). Oxford: Oxford University Press.
Saggar, S. (2003). Immigration and the Politics of Public Opinion. *The Political Quarterly, 74*(1), 178–194.
Schain, M. (2008). Commentary: Why Political Parties Matter. *Journal of European Public Policy, 15*(3), 465–470.
Schimmelfennig, F. (2001). The Community Trap: Liberal Norms, Rhetorical Action, and the Eastern Enlargement of the European Union. *International Organization, 55*(1), 47–80.
Schmidt, M. (1996). When Parties Matter: A Review of the Possibilities and Limits of Partisan Influence on Public Policy. *European Journal of Political Research, 30*(2), 155–183.
Seldon, A., Snowdon, P., & Collings, A. (2007). *Blair Unbound*. London: Simon and Schuster.
Shaw, E. (2002). *The Labour Party Since 1979: Crisis and Transformation*. London: Routledge.
Shi, T. (2008). British National Identity in the 21st Century. *Intercultural Communications Studies, 17*(1), 102–114.
Smith. (1990). *Labour Party Archive, NEC Annual Conference Report 1990*, 30/09/90–05/10/90. Labour History Archive Study Centre [LHASC].
Spehar, A., Bucken-Knapp, G., & Hinnfors, J. (2011). *Ideology and Entry Policy: Why Non-socialist Parties in Sweden Support Open Door Migration Policies.* Paper Presented at The Politics of Migration: Citizenship, Inclusion and Discourse in Europe, Glasgow Caledonian University, Scotland, November 10–12, 2011.
Stevenson, A. (2013, March 7). Interview with BNP "smasher" Margaret Hodge. politic.co.uk. Available from: http://www.politics.co.uk/comment-analysis/2013/03/07/interview-bnp-smasher-margaret-hodge. Accessed 6 Jan 2014.
Sun. (2001, March 7). Hague's Error. *Sun*.
Taggart, P. (1998). A Touchstone of Dissent: Euroscepticism in Contemporary Western European Party Systems. *European Journal of Political Research, 33*(3), 363–388.
Taylor, G. R. (1997). *Labour's Renewal: The Policy Review and Beyond*. London: Macmillan.
Thatcher, M. (1990, August 5). *Vision of Britain in Shaping the New Global Community*. Colorado: Aspen. Thatcher Archive: COI.

The Guardian. (2001, March 28). Speech Row for Tory Refuse to Sign Race Pledge. *Guardian*. Available from: http://www.theguardian.com/politics/2001/mar/28/election2001.uk. Accessed 13 Apr 2012.

Watson, M., & Hay, C. (2003). The Discourse of Globalisation and the Logic of No Alternative: Rendering the Contingent Necessary in the Political Economy of New Labour. *Policy and Politics, 31*(3), 289–305.

CHAPTER 6

Bringing the State Back In: Institutional Change and the Administrative Context

'It's in our DNA to look at things in a certain way'
(Interview with Civil Servant, 2012)

6.1 Introduction

The previous two chapters examined the role of organised interests and political parties in immigration policymaking and considered whether they were the key actors steering the managed migration agenda. In this chapter we turn to the administrative context and the policymaking process itself as an explanation for policy change.

The new institutionalist approach claims that political institutions can be autonomous, and it is these apolitical institutions that form immigration policy according to the interests of the state. Through processes of normalisation and socialisation, certain ideas, objectives and policy framings become embedded in these institutions, which serve as cognitive filters through which actors come to interpret their environment. In this way, ideas and framings become codified into institutional practice and 'take on a life of their own' (Berman 1998, 18) to the extent that they become resilient to change. Yet, through endogenous processes of institutional change, and/or exogenous political crises, historical institutionalism claims that such framings can be dislodged and replaced by a new set of objectives, ideas and framings.

Taking a disaggregated view of the state then—by examining departmental cultures, policy framings and institutional change—we turn to how the ideas of the modernised Labour Party were filtered through departmental agendas and changes to policymaking practices.

The chapter is divided into four sections. The first section describes the institutions and traditions that led to a stable and entrenched framing prior to 1997. The second section addresses the way in which Labour's third way ideology was exported and reflected in policymaking practices. The chapter then outlines how and why departments other than the Home Office became integrated into the policy network, in turn reflecting on how Labour's reforms to policymaking practices influenced immigration policy reform. We then turn to whether the initiative of evidence-based policymaking had an impact in the field of immigration and whether such evidence proved to be instrumental in the policy shift.

6.2 Departmentalism and Home Office Culture

To understand and explain policy change it is necessary to understand why policy remains static in the first instance. Policy change is often said to be incremental, piecemeal and gradual. In turn radical shifts in policy objectives and framings, such as the one under study here, are exceptional (Howlett et al. 2009). The typically gradual process of policy change is in part borne out of the fact that policy framings tend to be entrenched and institutionalised in government departments. Consequently policy often remains constant and stagnant for long periods, often for no reason other than institutional inertia.

The role of the 'British Political Tradition' and its tendency to direct government towards working within the existing status quo has long been noted as an explanation for policy stability in Britain (Blunkett and Richards 2011, 184). One such tradition is departmentalism, which involves a 'mix of political, policy and governmental pathologies' (Kavanagh and Richards 2001, 1), including a process where government departments have, over time, developed an organisational culture and a set of practices that are resistant to change. The culture of a department evolves from experience and acquired 'wisdom', such as which interest groups to consult with and the best way to effectively negotiate with the Treasury (Kavanagh and Richards 2001, 2). The history of the Whitehall model—with practices of ministerial accountability and performance and budgets being allocated according to departments—has meant that the civil service evolved around the notion of rigid departmental boundaries and entrenched silo mentalities (Pollitt 2003, 42; Richards and Smith

2005, 3; Bogdanor 2005, 4). The long held practices of protecting budgets and maintaining departmental 'turf'—where a department seeks to maintain or extend the range of responsibilities—have only exacerbated the isolation of departments and in turn disaggregated governing (Page 2005, 142).

Administrative processes framed around departments have cultivated bureaucratic politics then where political agents, such as officials, ministers and civil servants, 'see things differently from colleagues in other departments because their organizations have different objectives, ways of doing things, and because they have been socialized into thinking and acting in different ways' (Page 2005, 143). Therefore knowledge and the appropriate recourse to a policy issue are to some extent framed according to departmental thinking. Departmental knowledge and policy solutions become institutionalised and resilient to change. As one senior civil servant put it in an interview, 'it's in our DNA to look at things in a certain way'. A partial consequence of departmentalism then is that policies tend to remain within a static framing.

It is the way in which policy framings are embedded that helps to explain how new actors, from different departments, integrated into a policy network that had hitherto been confined, can bring new ideas, a new policy framing and in turn prompt policy change.

6.2.1 Home Office: 'A Culture of Caution'

Prior to the Labour administrations of 1997–2010 immigration policymaking was primarily dominated by the Home Office, which partly explains the relative consistency of immigration policy before Labour's reforms. While some administrative functions of immigration, such as the Work Permit scheme, were relegated to other departments, the politics and policymaking of immigration were monopolised by the Home Office. And like all other long-established Whitehall departments, the Home Office has acquired a certain culture, 'the axiom round which has been the attempt to find a balance between maintaining civic peace, while, at the same time, ensuring the liberty of the individual' (Richards and Smith 1997, 70).

What distinguishes the Home Office as exceptional in comparison to other government departments are the domestic and populist pressures it faces. It is, according to one former Home Secretary, 'a weird place. There is no sense or feel for the rhythm of politics, how the tide is flowing and the need to be ahead of things rather than trailing behind' (Blunkett 2006, 429). Because the Home Office was the dominant (and for a long time the only) social and domestic affairs department, it faces more media and public scrutiny than other departments, and consequently 'Home Office business is the staple for the populist press' (Interview with former Senior Official, 2012). Indeed most politically salient cross-cutting 'wicked issues' fall within the Home Office remit, and these issues resonate strongly with the public. These policy areas are predominantly related to negative and socially destructive issues, such as counter-terrorism, crime and policing and drugs policy. These are all areas dominated by the need for control and enforcement. Immigration policy was, and arguably remains, thus framed within the Home Office as an enforcement issue, similar to crime.

On account of the 'controlling' policy areas within the Home Office remit, the Department must be more reactive to media pressures, and the Home Office has thus developed a defensive and cautious culture, which feeds into policymaking. As former Home Secretary James Callaghan said of Home Office issues, 'a remote-controlled bomb is concealed in nearly every one' (Callaghan 1983, 10). As a result, 'the Home Office is politically driven, focused on short-term decision-making and tomorrow's headlines' (Interview with HO CV, 2012). In turn the jobs of Home Secretary and particularly Immigration Minister are deemed the most challenging in Whitehall. As former Home Secretary David Blunkett said of being Immigration Minister, 'it is a rotten job, and you are lucky if you get in and out quickly enough not to have been scarred' (Blunkett 2006, 712). With immigration being such a potentially salient issue, policymaking in the Home Office was, unsurprisingly, defensive and reactive leading to 'a focus on immediate problems and control rather than any strategic developments' (Interview with senior CV, 2012).

Prior to the immigration policy shift in the early 2000s, the Home Office had a long history of maintaining restrictive immigration policies and 'conceived itself as a rough, tough enforcement Department' (Interview with Tim Finch, 2011). The common culture binding the Home Office was an emphasis on casework and a feeling of what one interviewee descried as 'politically under suspicion'. A former communications director at the Refugee Council recalls an incident that somewhat encapsulates this culture:

> We were trying to persuade what became UKBA to get involved in refugee week which was very, very difficult. In the process of turning a lot of people

down for asylum, they actually granted asylum to quite a lot of people, and changed people's lives. But if you said to them "well that's part of your brief as much as enforcement", they looked at you as though you were a bit mad really. Because the whole ethos of the department was the "what we do is be horrible to people". (Interview 2011)

None of this is to imply that the Home Office had (or has) some ulterior political motive; rather the focus on enforcement and a defensive response to external pressures had become the norm. One former Home Secretary went so far as to suggest that the UKBA frontline staff had a 'sense of fatalism. They just picked up the dirty stick at every single point'. The fundamental Home Office objective was to 'keep the immigration systems tight' (Interview with HO CV). This fed down to the operational front-line staff and before 1997 this objective came to define all aspects of immigration policy. Asylum continued to be couched in a control frame, and components of this culture certainly remain in place today. But in work-related immigration, under Labour, this framing was somewhat reshaped in the early 2000s.

6.3 The Politics of New Labour: Changing Institutional Settings

All too often governments in the past have tried to slice problems up into separate packages – as if you could fix an estate by just painting the houses rather than tackling the lack of jobs or the level of crime....Joined up problems demand joined up solutions... government itself has to change if it is to be the solution rather than, as is sometimes the case, being part of the problem (Tony Blair 1997)

This government has given a clear commitment that we will be guided not by dogma but by an open-minded approach to understanding what works and why. This is central to our agenda for modernising government: using information and knowledge much more effectively and creatively at the heart of policy-making and policy delivery (David Blunkett, Speech to the ESRC, 2000)

Upon electoral victory in 1997, the Labour government identified departmentalism as a fundamental barrier to effective policymaking. Frustrated by what Labour elites considered an idle civil service, Blair was particularly keen to break with the past and reinvigorate the machinery of government: 'The problem with them, as I indicated at the beginning, was inertia. They tended to surrender, whether to vested interests, to the status quo or to the safest way to manage things—which all meant: to do nothing' (Blair 2010, 205). In response, the government proposed a renovation of the machinery of government with the aim of coordinated policymaking, otherwise known as the joined-up government (JUG) strategy. In conjunction with the JUG agenda, in an attempt to depoliticise and rationalise policy (Burnham 2001), the Labour government endeavoured to practise evidence-based policymaking (EBP). Both initiatives were symptomatic of Labour's third-way framework, in particular the Party's sought out objective to modernise policy on a 'what works' basis.

The JUG strategy, first articulated in the White Paper *Modernising Government* (1999a), derived from the belief that policy could not be effectively delivered through the separate and isolated activities of individual organisations. Specifically, 'wicked issues'—issues that had long cut across departmental boundaries with no sufficient solution—were seen to be lacking in effective policy (Mandleson 1996; Bogdanor 2005). One such area was, and remains, immigration. The mechanisms to deal with conflict between departments were also viewed as weak and ineffective (Clark 2002, 108; Ling 2002, 612).

The drive for holistic policymaking was due to: (1) the need to address policy issues that cut across departmental boundaries; (2) 'spillover effects' where one part of government fails to take account of its impact on other parts of government; (3) a problem of organisation and integration such as 'how to align incentives, cultures, and structures of authority to fit critical tasks that cut across organisational boundaries' (Moseley 2009, 2; Mulgan 2005, 175). The focal point of the JUG strategy was the concept of partnership, the aim being to co-ordinate consensual policy through multiple departments without removing departmental boundaries (Pollitt 2003; Clark 2002, 107).

Both scholars and policymakers themselves have disputed the novelty and effectiveness of the JUG strategy. Whilst, in theory, JUG was an attempt to make policymaking more holistic, the programme involved costs, as the government conceded, 'Cross-cutting approaches are no panacea' (Cabinet Office 2000, para. 4.5). In general cross-government working poses greater risks of failure in delivery due to unclear lines of accountability, greater risk of communication failures and difficulties in evaluating policies because of cross-government ownership (Pollitt 2003).

Also part of Labour's efforts to modernise policymaking was a drive for evidence-based policy. Evidence-based policymaking derived from Labour's interpretation of modernity, structured around efficiency, knowledge management and technological and scientific progressivism (Finlayson 2003, 67). The initiative was an attempt to improve policymaking so that policy was depoliticised, 'forward looking' and 'shaped by the evidence rather than short-term pressures' (Burnham 2001):

> ...policy decisions should be based on sound evidence. The raw ingredient of evidence is information. Good quality policy making depends on high quality information, derived from a variety of sources – expert knowledge; existing domestic and international research; existing statistics; stakeholder consultation; evaluation of previous policies. (Cabinet Office 1999a, 31)

Labour sought out a process of policymaking where evidence would be an integral component, and policy would be based on 'what works'. Promoted in numerous Cabinet papers (Cabinet Office 1999a, b, 2001; National Audit Office, 2001), the ambition to integrate evidence and, in turn, knowledge into the policy wheel was a clear objective of the Labour governments, especially in Labour's first term of office. The rhetoric and practice of JUG wavered in Labour's second term and had been abandoned by Labour's final term (see Caulkin 2006). But in the first term, when EBP was thriving, evidence-based policymaking meant expanded research budgets across Whitehall, an emphasis on evaluation as a widely accepted part of policymaking as well as input from outside experts through secondments (Wells 2007, 27).

Yet much like the JUG strategy, in practice evidence-based policymaking was problematic. In addition to wider debates about what constitutes evidence in the first instance, 'the dynamics of policymaking are deeply affected by institutional, professional and cultural factors' (Head 2010, 80); thus the notion that evidence can be simply and objectively introduced into the policy wheel has always been ambiguous. Other limitations of evidence-based policymaking include: how to translate findings into policy, a tendency among political elites to use evidence to legitimise positions, thus discrediting objectivity (Boswell 2009) and a lack of information needed to gather evidence. In other words 'the more we discover about social issues the more we are likely to become aware of the gaps in our knowledge' (Head 2010, 80).

It is worth noting that the JUG model and evidence-based policymaking were not designed to pluralise power in Whitehall. On the contrary, Blair's objective with the JUG strategy was to establish a clear and coordinated agenda accountable to the core executive, something he readily admitted: 'One thing I do say though very strongly is that I make no apology for having a strong centre' (Liaison Committee 2001/2002, para. 5; Richards and Smith 2007, 331). For example, the creation of units such as the Performance and Innovation Unit (PIU) and expanding the size of the Policy Unit, whilst products of both the evidence-based policy and JUG agendas, were intended to allow more central steering and were thus directly accountable to the Prime Minister's Office.

While the successes of both the JUG initiative and evidence-based policy model are contestable, the empirical findings demonstrate that these adaptations to policymaking practices influenced economic immigration policy. These reforms to the machinery of government help provide an explanation for not so much *why*, but *how* the policy shifted.

6.4 From Control Thinking to Economic Thinking: Immigration Policy and Joined-Up Government

Rather than a central government strategy for an expansive immigration programme, policy change was a result of a series of policy reforms that stemmed from departments other than the Home Office. Each policy scheme or reform had its own drivers, couched in the departmental culture and framing in which it originated, which were deemed politically non-contentious and unproblematic at the time. Combined, these policy schemes had a liberalising effect on immigration policy, and in this way the overarching shift was almost a by-product of other policy agendas (Wright 2010). Managed migration was a label devised to provide a cohesive narrative to policy changes that were already under way. Thus the policy change was not so much steered as it was piecemeal and unintentional.

Given that 'immigration was a classic example of a topic which needed some joining up. It clearly cut across so many different departments—and there was an obvious risk of contradictory decisions', there was a perceived need for JUG in this area (Interview with former head of Strategy Unit, 2012). As a result, the policy network in terms of state actors expanded under the Labour administrations. The inclusion of new departments into a previously confined policy network (as well non-governmental actors seconded in) broke the Home Office monopoly on immigration policy-making and brought different targets, objectives and a new framing to the policy area. With strong economic conditions, and an especially powerful Chancellor, the Treasury who had previously argued for the case of expanding migration routes had found a governing party sympathetic to their arguments. The Treasury's early liberalising schemes combined with involvement from other business-orientated departments imported a somewhat rationalistic framing to immigration for the first time, thus transforming the norms of immigration policy in the Home Office from a control focus to an economic policy objective.

6.4.1 The Treasury

Outside the Home Office, the most influential department involved in economic immigration policy was the Treasury, which, under the Labour administrations, had a strong influence on the DWP and the DTI and as a Department 'was extraordinarily confident. This was Brown's Treasury and people used to refer to him as PM domestic' (Interview with former senior CV, 2012).

The Treasury is especially powerful because it is 'the department that raises questions about what is in the taxpayer's interest…The Treasury has to ensure that the enthusiasms of other departments for their various policies are optimised and balanced' (Chapman 2002, 164). The power the Treasury exercises is achieved through 'persuasion and influence' (Chapman 2002, 160) and officials in other departments are said to have a 'wary respect for Treasury staff' (Chapman 2002, 160).

The complex partnership between Tony Blair and Gordon Brown is well known. Brown secured an active role in domestic policy in return for not challenging Blair's bid for party leadership in 1994, otherwise known as the 'Blair-Brown deal'. In return Blair had to gratify and accommodate Brown, and Brown therefore had a great deal of autonomy outside the typical remit of a chancellor (Heffernan 2006, 594). It is no surprise then that the Treasury under Brown's leadership has been described as 'a great crag standing in the way of a thoroughly monocratic government' (Hennessy 2002, 21). While the relationship between chancellors and prime ministers has always been consequential, 'the Blair-Brown partnership [was] far more significant than any such previous relationship…

Brown held a special position that coloured the structure and operation of the Blair governments' (Burch and Holliday 2004, 19).

Besides the chancellor's formal power of managing the public purse, Brown amended the Treasury's role in Whitehall management to further consolidate his power, including establishing Public Service Agreements (PSAs) and performance targets, which allowed Brown to have direct control over not only how much, but also what departments could spend their budgets on (Fawcett and Rhodes 2007, 95). With the unique partnership between Blair and Brown dictating institutional reforms, Labour's first two terms were characterised by a strengthening of the core executive and an especially powerful Treasury. Crucial to the Labour Party's economic policy was labour market flexibility. Increasing the supply by expanding immigration flows as a remedy for such flexibility was pursued through the Treasury's agenda.

The Treasury were the innovators for the new approach to immigration policy, arguably creating a specific sub-policy of economic immigration altogether. As one interviewee put it, 'the thinking about what the strategy should be was coming from BIS and the Treasury, and the immigration department was just playing catch up if you like' (Interview with Senior CV, 2012). Indeed, interviewees claimed that the overall shift was due to the 'Treasury pushing the Home Office to do more and more' (Interview with Senior CV, 2012).

It was the Treasury that first set up cross-departmental groups to develop the managed migration narrative in 2003, and it was the Treasury that established the Innovators Scheme in 2000 and the HSMP in 2001, years before the Home Office introduced the concept of 'managed migration' (Home Office 2002). These schemes were driven by a supply-side logic, encouraging high-skilled immigration to fill skill shortages as part of a broader economic growth strategy, as outlined in the Treasury's 2000 budget:

> Failing to fill [job] vacancies with skilled workers will retard productivity and growth and mean fewer employment opportunities in the longer term, access to skilled people from overseas is part of the answer. Equally important is to enhance the UK's image as an attractive location for talented overseas students and entrepreneurs. (HM Treasury 2000a, 2.76–3.77)

The impetus for the schemes derived from the Treasury's productivity and growth agenda, which, following analysis on Britain's 'productivity

6.4 FROM CONTROL THINKING TO ECONOMIC THINKING: IMMIGRATION... 171

gap', concluded that 'to increase productivity growth, the UK has to increase its investment in all three areas of physical capital, human capital and technological progress' (HM Treasury 2000b, 39). The Treasury therefore prioritised increasing investment in physical and human capital, strengthening competition and encouraging entrepreneurs and entrepreneurship (HM Treasury 2000b, 40). Consequently, the Treasury's 2001–2004 spending review—and therefore the joint PSA target between the Treasury and the Home Office—outlined the principal objectives of immigration policy to be 'in the interests of social stability and economic growth' (HM Treasury 2000c). Further PSA targets likewise alluded to economic growth being an objective of immigration policy, albeit vaguely stated, such as PSA Five under the 2004 Home Office spending review, where the objective to be met (aside from tackling unfounded asylum claims) was to 'promote controlled legal migration' (Cabinet Office 2006, 9). Whilst the phrasing was amended, the 2007 PSA departmental strategic objective was likewise to 'strengthen our borders, fast track asylum decisions, ensure and enforce compliance with our immigration laws and boost Britain's economy' (Home Office 2007).

Many of the early policy changes from the Treasury were not policy as such, with the only public announcement of schemes (such as the HSMP) in pre-budget reports. For example, liberalising policies—such as widening the definition of 'skilled' to allow students to work in the UK post-study—were part of a small section entitled 'tackling skills shortages' in a pre-budget report (HM Treasury 2000c). Conscious of the politics of immigration and typical of the Treasury's 'semi-independent relationship with parliament' (Heffernan 2006, 591), some interviewees suggested that the Treasury were deliberately restrained in publicising the reforms in order to 'get stuff through' without 'getting a great deal of political attention'.

> It was a strategy but apart from Roche's speech it wasn't a well-publicised strategy and in terms of immigration, it was happening in a corner of immigration. (Interview with senior HO CV, 2012)

The Treasury were particularly interested in the effect of immigration on public finances, increasing its tax base, and most importantly the positive impact on GDP. The Treasury cited the economic benefits of immigration earlier than the general political rhetoric of managed migration, claiming in 2002 that economic growth had been boosted by a quarter of a point because of immigration (Schifferes 2002). A former civil servant explains the Treasury's sudden interest in immigration:

> The Treasury view was that it made sense that there were a significant numbers of people around the world whose skills were so strong that they felt that we self-evidently wanted them in our country and therefore they should be free to enter our country to seek work. And so the Treasury worked with the Home Office to...initially persuade them to have a programme like this, and then ultimately to implement that. (Interview with former CV, 2012)

Many interviewees commented that there was 'a bias [in the Treasury] which were in favour of open labour markets' and that 'most of the economists had a strong prior belief in what the answer should be' (Interview with former Director of Strategy Unit, 2012). While the Treasury had argued the case for expanding economic migration routes under preceding governments, under the Labour government 'it found a set of ministers in the late 1990s and early 2000s who were sympathetic to this' (Interview with former SpAd, 2014). Some interviewees suggested that expansive immigration policies were crucial to the Treasury's economic model, although this was never confirmed in any official publication:

> The Treasury's economic models explicitly depended on migration holding down wages. If you look at the Treasury economic models especially from mid 2000s onwards and possibly before, there was an assumption that migration would reduce wage levels and help improve growth. So there was this strand for New Labour you could say, of deliberately using migration to depress wages. I am told that the Treasury model explicitly assumed this. (Interview with former minister, 2012)

There was speculation among some interviewees that the Treasury's motivation for expanding economic immigration derived from the inference that labour flexibility lowers inflation rates. The only indication that this was an explicit motivation was in one Treasury report, which concluded that migrants, in meeting skill and labour shortages, can 'ease inflationary pressures' (HM Treasury 2004, 4.60). While lowering inflation was not categorically expressed as a principal driver for expanding immigration, given the importance Labour placed on achieving counter-inflationary credibility, there may have been an implicit assumption that increasing the supply of the labour force could ease inflation. This may have figured as a secondary motivation then. As a former MAC member speculated in an interview '[lowering inflation] may have been part of the rhetoric or discourse of increases in GDP lowers inflation. But I don't think the Bank of England went down to Downing Street and said we need to do this because it lowers inflation'.

The Treasury's legitimisation of expansive immigration policies was that immigration boosted economic growth by one and a half per cent (Blair 2004). This was derived from the Treasury's estimations of a 2.5 per cent trend growth in the economy, of which 2.0 per cent came from productivity increases and 0.6 per cent from increases in the population of working age (HM Treasury 2002). Further figures produced by the Treasury that were trailed by ministers (Smith 2007; Byrne 2008a) concluded that immigration has a positive impact on the economy worth approximately £6 billion; this figure was based on Treasury calculations that immigration contributed 15–20 per cent to output growth during the period 2001–2006. Further research on the fiscal impact of immigration conducted by the Home Office and the PIU concluded that migration generates a 'net benefit of 2.5 billion to the economy, over and above what they consume' (Gott and Johnston 2002). This latter figure was parroted in numerous speeches, press releases and parliamentary debates (Blunkett 2002; Hughes Hansard 28 October 2002, vol 391). However, the Treasury's much-quoted figure (and indeed the Home Office figure on fiscal impact) was not based on an especially high-level analysis:

> That treasury calculation is very simple, it's not based on any ecometrics, it's based on looking at that migrant share as a population. You could do it on the back of an envelope, it wasn't based on anything sophisticated at all and I don't think anything else sophisticated was going on in government elsewhere. (Interview with CV, 2012)

By Labour's third term, the figure and methodology of measuring the economic impact of immigration were a contested issue, leading to a House of Lords inquiry in 2008. The inquiry concluded that there is,

> No evidence for the argument, made by the Government, business and many others, that net immigration—immigration minus emigration—generates significant economic benefits for the existing UK population. Overall GDP, which the Government has persistently emphasised, is an irrelevant and misleading criterion for assessing the economic impacts of immigration on the UK. (House of Lords 2008, 1)

While the Treasury played a central role in the policy shift, the Treasury's interest and resources into the issue should not be overstated. The Treasury did not invest substantially into research on the economics of immigration and actual work on immigration took up only half a person's job (Interview with CV, 2012). The Treasury's authoritative voice on the economics of immigration was more a result of inertia from other departments, especially the Home Office: 'it's not that the Treasury were doing loads of work on it but the most quoted statistic originated from the Treasury and probably no one in government was doing serious thinking about it at that time' (Interview with former CV, 2012). The Treasury was important in pushing for policy reforms, yet expanding immigration was not a critical issue to the Department, or one worth investing too many resources into. However, because of the lack of direction or lead from the Home Office, the Treasury's role was behind the scenes but nonetheless imperative and the impetus for policy change:

> Other than saying things in the budget and the PBR, the Treasury never made a public announcement in any other context, it wasn't their job to front the policy, it was the Home Office's job, which is part of the reason the Treasury and Home Office were a team. There isn't a stakeholder management function in the Treasury. (Interview with former civil servant, 2012)

As a result of the Treasury's involvement, the Home Office brought economists into their policymaking teams for the first time. The powerful influence of the Treasury pushed economic immigration in an expansive direction and, 'because of the relationship with the City of London, they [Treasury] were strongly influenced by the London view of the British economy, which was that inflows both on the high end and the low end had to be critical to growth' (Interview with former Director of Strategy Unit, 2012). The Treasury's early involvement and powerful position effectively led to an economically rationalistic framing on immigration being imported into the Home Office for the first time. The norms of a control policy frame began to shift to an economic frame and a policy based on economic utilitarian arguments (Balch 2010).

6.4.2 Business-Orientated Departments

While the Treasury provided the impetus for policy change, other business-orientated departments were also more involved in immigration policy under Labour, including the Department for Education and Employment (DfEE) and the Department for Trade and Industry (DTI). The extension of 'turf' to these departments was in part a consequence of the pursuit of JUG. The channelling of JUG was also reflected in the significant increase in cabinet sub-committees, interdepartmental taskforces (especially in the design of the PBS) and the abovementioned cross-cutting PSA targets (HM Government 2009; Spencer 2011, 19). While the JUG strategy was not wholly successful, the Home Office monopoly on immigration was clearly fractured, and immigration was consequently no longer framed entirely around control and restriction.

It was in fact the DTI that first suggested liberalising economic immigration policies back in 1998, as part of a wider economic growth strategy, in an effort to move away from 'the interventionist policies of the past' and 'instead make markets work better through collaboration between companies as well as competition' (Mandelson 2010, 265), as outlined in their 1998 White Paper:

> It is important to attract bright people with scarce skills to work for UK businesses and to set up businesses of their own which create jobs. The Government is already taking action to ensure that the process of obtaining work permits is as smooth and simple as possible. Government will examine whether there is scope for lowering barriers that prevent entrepreneurs or skilled professionals from coming into or remaining in the UK. (DTI 1998, 3.24)

In conjunction with the Treasury, the DTI's objective was to plug skills shortages by attracting high-skilled immigrants, and to fulfil this objective the DTI claimed that 'this requires a positive attitude to immigration' (DTI 1998, 3.24). The genesis of this shift in policy framing can therefore be seen as far back as 1998, yet the objective stemmed from a department with no management of, or previous remit, on immigration.

That certain elements of immigration policy were delegated to other departments is significant in this narrative of policy change. The work permit system had previously been based in DfEE before 2001, and changes to the work permit system can be seen as a result of these functions being placed in a business-friendly department (Düvell and Jordan 2003). For example, the CBI held a conference on overseas labour (as economic immigration was then referred to) in the late 1990s where Margaret Hodge (then junior employment minister) made a presentation effectively asking what employers wanted regarding overseas labour and reducing red tape (Interview with Sarah Spencer 2011). In contrast to the Home Office, which had never comprehensively conducted stakeholder engagement, in terms of immigration policy at least, DfEE organisational culture was very different because stakeholder management was common practice. Margaret Hodge described the Department in an interview:

> I think we were [DfEE] very open in our policymaking. So you'd always talk to your stakeholders. Policymaking was a collaborative effort with your stakeholders, always. I think we were a department that was open to change. You're always trying to raise standards, you're always trying to get better, so you're open to innovation. We brought people in from outside much better than other departments. The civil service is a pretty closed institution, but our stakeholders are out there and they are a particularly articulate lot.

Other interviewees commented that the DfEE was more 'progressive' than the Home Office in that there tended to be multi-disciplinary teams (including researchers, economists and operational managers) working in collaboration. Another former civil servant commented that DfEE staff saw 'their mission in life to find good people and let them come into our country, whereas the Home Office saw their mission as to stop letting the bad people into our country. They had a very different way of thinking about their job' (Interview with former CV, 2012). The easing of work permit criteria and the rapid and uncontested way in which this was conducted in the late 1990s is a reflection of such functions being placed in DfEE.

Aside from the work permit scheme (until 2001), the DfEE also owned the 'Prime Minister's Initiative on International Education' (PMI1 and PMI2) although this was a personal pledge made by Prime Minister Blair. The PMI aimed (and succeeded) to double the number of international students coming to study in Britain. The drive to recruit international students was hardly seen as part of immigration policy, and in contrast to recent debates about student migration, international students were seen as an entirely positive migration stream. Employment minister at the time Margaret Hodge commented that the push to recruit international students and build links with foreign universities came from 'Tony Blair, but it also came from the universities themselves, it was encouraged by all of us, encouraged by Gordon Brown. I don't think it was a contentious issue'. Moreover, asked whether international students were seen as part of the managed migration strategy Hodge responded 'it was just seen as extra money for universities' (Interview with Margaret Hodge, 2012). Similarly, in response to the same question former Secretary of State for Innovation, Universities and Skills John Denham replied:

> No, not at all, not in the way they are now. They [international students] were seen as an almost wholly beneficial economic good to the country and the university system, at a time when public funding for higher education was going to stop growing.... I mean the fact is when we extended the right to work [to international students], there was no questions raised about that in the Home Office at all.

The PMI1 scheme began in 1999, years before the managed migration strategy was outlined. Much like the Treasury's early schemes, the PMI initiative was not a policy driven by a desire to liberalise immigration as such, but was rather seen as a non-contentious way to generate money for universities. The initiative, which led to student immigration being the largest immigration stream to Britain, was a by-product of the higher education policy agenda. The impact of migration on skills level was also a consistent interest for DfEE, as former Home Secretary Charles Clarke commented in an interview 'Skills is at the core of the whole debate…the interrelationship between the skills agenda in education and the migration agenda is very substantial'.

6.4.3 Foreign and Commonwealth Office

Another important department in this narrative of policy change was the Foreign and Commonwealth Office (FCO). Migration policy has long been an issue of contention between the Home Office and the FCO. This dispute stems back to the 1970s: 'there is a longstanding and somewhat futile disagreement between the Home Office and the FCO on whether there exists a "finite" pool of potential immigrants waiting to come in or an "infinite" ocean that will never be exhausted' (TNA FCO 50/585 24 May 1976). Ownership of migration functions in particular has been a persistent cause of friction between the departments (TNA FCO 50//532 August 1975). Over time the FCO has lost, somewhat willingly, much of its remit on immigration. Visa processing for example has been consolidated in the Home Office and the FCO's remit for migration policy is now limited to the operational role of overseas embassies.

The FCO therefore had a minor and traditional agenda under Labour to keep the status quo of using migration as a 'lever in international relations' for negotiations (Interview with FCO CV, 2012). The Department purportedly has an 'insular culture', however, which tends to 'struggle with the cross-Whitehall aspect' and is 'insufficiently politically aware' in the sense that the department fails to 'be thinking about wider government objectives' (Interview with CV, 2012).

While the FCO gradually lost its authority on immigration policy, conversely the most controversial and expansive immigration policy was co-led by the FCO – the A8 decision in 2004. While the Home Office had departmental responsibility for the decision, and it was legitimised by the Home Secretary with reference to the managed migration agenda, this decision was, at the very least partly, driven by foreign policy considerations (Interview with Senior CV, 2012; Interview with Matt Cavangh, 2012; Interview with Ed Owens, 2011; Interview with Dennis MacSchane, 2012; Interview with former official, 2012).

It was Foreign Secretary Jack Straw who announced the A8 decision in December 2002 and it was Straw and Blunkett who predominantly led the decision, with Blair 'involved at the last minute when it became heated' (Interview with Blunkett, 2012). A number of interviewees commented that the decision was 'rather rushed' because Blair wanted to announce it at the Copenhagen Council in December 2002 (Interview with former SpAd, 2011). Therefore 'it was one of those things where a quick letter goes round Whitehall and it's sort of "is there any objections?", that's how it worked' (Interview with former SpAd, 2011). The interest in Central and Eastern Europe (CEE) accession was principally due to trading ties and a foreign policy interest in forging alliances with CEE states at the EU level. Labour MP Keith Vaz explicitly divulged this strategy in a debate, claiming that to modernise Europe 'we need not only to harness the enthusiasm of the new members but to earn their votes around the summit table' (Hansard, 2 March 2004, cols 870). CEE states were thus regarded as allies of British interests, as one FCO civil servant deliberated:

> The Slovaks and the Poles were useful and saw things in the same way as Britain. For instance our approach to the EU is to think what is it that the EU does that helps us to take forward British objectives, so it's not integration for its own sake, and that's where the Poles and Slovak saw it at the time.

As Home Secretary at the time David Blunkett put it 'because of our [Britain's] links with Poland, the Polish diaspora in Britain offer quite a good counterweight to the German/French axis without making it aggressive' (Interview with Blunkett, 2012).

Labour's Europhile approach is also relevant to the decision. A number of interviewees including former European Minister Denis MacShane maintained that a more Eurosceptic Labour Party would have likely made a different decision (Interview with Dennis MacShane, 2012). In turn the Commission was particularly pleased with the government's decision to embrace enlargement and free movement (Interview with FCO civil servant, 2012) because 'diplomatically the fact that Britain was prepared to do this, to show solidarity with CEE, was seen as a real value' (Interview with former SpAd, 2011).

In contrast to geopolitical motivations, former Home Secretary David Blunkett claims the decision was made to reduce irregular immigration. Of those A8 migrants who registered on the WRS, Blunkett claimed that 40 per cent were estimated to have been living in the UK illegally pre 2004 (Interview with David Blunkett, 2012). In other words to regularise A8 migrants who were already working in the UK illegally:

> The decision was are we going to have illegal working and no payment of national insurance tax or are we going to try and make this legal, and getting people to register? Make it worthwhile people registering because you build up entitlements very quickly, and therefore you were going to be entitled to the health service and benefits and all the rest of it. (Interview with Blunkett, 2012)

According to Blunkett the liberalising effect of the A8 decision was, conversely, part of the control agenda. This was premised on the 'Blunkett equation' outlined in Chap. 4, that more 'good' immigration equals less 'bad' immigration (Balch 2010, 131). Blunkett claimed that Foreign Secretary Jack Straw was initially a 'great enthusiast for the expansion and openness of Europe' but following Conservative leader Michael Howard's attacks on the decision (discussed in Chap. 5) 'with the political heat turned up he changed his mind' (Blunkett 2006, 593). Having said this, a former official commented that the Home Office were relatively uninvolved in this decision but that regularising was an attractive advantage, as it would allow the Home Office to concentrate enforcement efforts on irregular immigration.

Jack Straw similarly alluded to the 'Blunkett equation', claiming that the decision would allow the government 'to focus resources on the real immigration problems' (FCO 2002). However, all interviewees commented that the decision would not have been made if they knew the numbers involved. In other words, the great migration wave from the CEE was certainly not anticipated:

> It was done on the expectation that the numbers wouldn't be great. It definitely wasn't a policy to actively go after Eastern European migrants. I mean the objective was not to draw on lots of Eastern European labour at all; it was simply that there didn't appear to be any reasons for controlling access to our labour market because we didn't expect the numbers. There didn't appear to be any big problems with it. There were Home Office officials responsible for dealing with those here on an irregular basis and deportation. Their view was if people can travel to the UK they can come here without needing visas, therefore it is incredibly difficult to police their access to the labour market and if you can't remove them there's no point. It's far better for them to come here to work. So paradoxically the people in the Home Office who were responsible for immigration control were in favour of it because it allowed them to concentrate on the cases they cared most about which was failed asylum seekers. (Interview with former SpAd, 2014)

While predictions suggested that A8 flows would be approximately 5,000–13,000 per year (Dustmann et al. 2005; Bauer and Zimmerman 1999), the decision turned out to be over 20 times the upper end of this estimate and between May 2004 and September 2005 293,000 A8 migrants entered Britain (Gilpin et al. 2006, 13; OECD 2009). However, the prediction was based on the assumption that other member states would also open their labour markets to A8 nationals; in particular it was assumed that Germany would not place transitional controls (Interview with civil servant, 2012). Whilst Britain confirmed the decision in late 2002—although some interviewees alluded to the decision having been effectively made when Blair entered office in 1997—most other Member States did not make the decision to impose transitional controls until early 2004. According to Blunkett (2006, 599) such decisions were taken in France and Germany because 'virtually every middle-class family had someone working for them who was a clandestine, so it was better to pretend that they were stopping people from working while they carried on employing them'.

Interviewees varied in their responses as to the primary reasons for the A8 decision and given that the foreign policymaking process remains the most secretive in government (Williams 2004, 911), it is difficult to definitively conclude the principal motivation. Although the rhetoric of managed migration had begun by 2004 and labour market shortages no doubt played a role in this decision, it seems this was a policy informed by Britain's broader foreign policy agenda to 'get more pull with the Poles within Eastern Europe. It was a bid for diplomacy!' (Interview with John Denham, 2012). With no capacity to deport A8 migrants if transitional controls had been placed, the Home Office view was that there was 'no point' in doing so and that it would be 'far better for them to come here to work' (Interview with Pearce, 2014). The labour market and economic conditions were right, alleviating irregular immigration by regularising Eastern Europeans was certainly an advantage, and the geopolitical gains were abundant. In this sense the decision was a triple win for the government:

> There was a belief by key decision makers that free movement was a fundamental part of the European project. Let's not forget the politics of this, Britain was a key advocate of EU expansion, it had key friendships in Poland and the Czech Republic which it wanted to maintain, and it wanted to be seen as living up to the spirit of EU enlargement, so it wasn't a crude what does the labour market need right now. (Interview with Danny Sriskandarajah, 2012)

Thus one of the largest immigration flows in Britain was not altogether an immigration policy; the decision was driven by a myriad of reasons, which were beneficial for both the Home Office and the FCO agendas. This was in part a foreign policy with unexpected ramifications for immigration policy, akin to the 1948 British Nationality Act (Hansen 2000). Much like the PMI scheme and the Treasury's early schemes, the decision was not considered an immigration policy per se; rather it was another by-product policy that had a liberalising effect on immigration.

6.4.4 Communities and Health

The hiving off of immigration functions from the Home Office also created closer collaboration between departments, such as integration and community cohesion policy being transferred from the Home Office to the newly established Department for Communities and Local Government (DCLG) in 2006. The DCLG had had some influence on immigration policy, such as devising the Workers Registration Scheme and chairing the Migrant Impacts Forum (MIF). However, in practice the links between national policy and local policy impacts were not especially strong. Often involvement between the Home Office and the DCLG on immigration was 'on a more issue by issue basis' (Interview with senior CV, 2012). In comparison to current immigration policymaking practices 'before 2004 the localised impact of immigration weren't felt as much so DCLG weren't players in quite the same way' (Interview with former SpAd, 2014). The Labour government's failure to accommodate or plan for the local impacts of concentrated immigration in specific areas has been one of the major criticisms of Labour's immigration policy. It could be speculated that this was caused by the poor communication or lack of strategic partnership between the Home Office and the DCLG.

The Department of Health continued to have a vested interest in foreign labour, especially to fill skill shortages in the National Health Service (NHS). Following various enquiries by the United Nations (2001), research was commissioned into whether immigration could be used to remedy Britain's 'demographic time-bomb', and there was a general debate both within the government and across the EU on whether this was a plausible solution (Home Office 2002; Government Actuary Department 2002; HM Treasury 2003, 3.103). However, there is no evidence that the Department of Health were involved in these enquiries, and various bodies concluded that immigration 'is not a long term solution to the "problem" of population ageing' (Shaw 2001; OECD 2001; Council of Europe 2000). Expanding immigration policies was not then a consequence of a concern over a demographic time bomb.

6.4.5 Departmental Tensions

Whilst government departments became involved in shaping and devising immigration policy, the relationships between departments were not necessarily harmonious, as many civil servants commented 'there were big tensions between the Home Office and the economic departments, precisely because the Home Office was being looked to, perhaps unrealistically, to provide control and protection' whilst concordantly pursuing an economically liberal policy (Interview with senior CV, 2012). The Home Office had developed a reputation from the business departments as being 'economically illiterate and only concerned about clamping down on things' (Interview with former official, 2012).

Similarly IND frontline staffs were not necessarily sympathetic to the new policy reforms and the tension between national policies on the one hand, and the realities of policy implementation on the other, came to the fore:

> Day-to-day it's a story of quite a lot of tensions in which the people who work in the organisation are quite crucial in that out on the front line they don't really buy into the managed migration strategy, they don't see the higher-level policy objectives. They had a well-embedded operational culture, which meant that they were not prepared to give effect to policy. And the policy was not sufficiently organisationally embedded. (Interview with former CV, 2012)

Another interviewee commented that there was a split between 'broadly liberal policy people' and 'more hard-end operations' people (Interview with former SpAd, 2014). The conflict between and within departments over immigration policy stems from the contradictory agendas combined with the contentious politics of immigration, making JUG working very difficult, as former Director of the Strategy Unit commented:

> I think it [immigration] was probably harder than most topics to deal with because the contradictions were more deeply rooted in the political stance of the government and to some extent unresolved... the public too had contradictory and perhaps incompatible desires for getting all the benefits of openness and none of the costs. That was what many people wanted, but it is very difficult to achieve. (Interview, 2012)

Despite these conflicts and difficulties in communication, the expansion of departments other than the Home Office claiming turf on work-related immigration policy changed both the practice of immigration policymaking and the framing and policy solutions from a control to an economic frame. This extension of turf was a result of two concordant changes in the administrative context: a mission for JUG and more importantly a particularly powerful Treasury initially pressurising the Home Office.

Where once policymaking was an insular affair monopolised by the Home Office and couched in a control frame, by 2004 when a review of the immigration system was conducted (eventually leading to the PBS) the

Home Office were conducting cross-government reports and cross Home Office taskforces, collaborating with caseworkers, enforcement people, visas, analysts, operational people, policy people, and economists.

There was an active move then to engage both across departments and within the Home Office in the early 2000s, and the influence of other departments engendered a slight cultural change in the Home Office. Collaborative working, especially across the Home Office, on economic immigration became commonplace and remains practice today, suggesting a lock-in effect of institutional practices. Before this period there were no economists looking at immigration in the Home Office and as one interviewee suggested 'I think that probably tells you something about where the priorities of the Home Office were at the time' (Interview with CV, 2012).

Prior to the Labour governments of 1997–2010 immigration policy had its own enclave. But with the influence of economically geared departments, especially the Treasury, the Home Office tried to understand how immigration could be made to fit with the wider policy objectives. The business-orientated departments played a significant role in importing the economically liberal ideas that were entrenched in their departmental cultures. This somewhat modified Home Office policymaking practices in the realm of economic immigration policy, a process symptomatic of 'institutional conversion' (Béland 2005; Thelen 2004). The Home Office was never charged with looking through the 'economic lens' before, but the framing of immigration from that to be restricted to an economic growth strategy was a result of integrating ideas from business departments:

> They [the Home Office] had spent 40 years thinking their job is to keep the doors totally shut, so getting them to think differently was really, really challenging and it was a really big culture change for them…By the time you get to 2005, the Home Office don't need that kind of pressure anymore, it's now starting to begin thinking about this stuff itself, even by 2002/03, it's beginning to embrace this stuff. (Interview with former CV, 2012)

Indeed, some interviewees suggested between 2000 and 2005 there had been a discernible shift or a 'period of enlightenment' in the Home Office, from policymaking being conducted in a closed manner to being a collaborative effort between government departments and stakeholders. Interviewees commented that by 2007 it was harder to maintain relationships as the Home Office was undergoing a crisis engendered by a failure of cross-government working. This was acute because the politics of immigration were so difficult it was hard to establish clear cross-department ownership.

The administrative context is thus pertinent to explaining the policy shift because the overarching 'third-order change' (Hall 1993) was ultimately a result of a series of by-products from other departments (Wright 2010, 177). These were not designed or conceived as one homogenous unified policy. Rather, these policies were spill-overs from other departmental agendas, leading to an immigration policy that was, somewhat ironically, anything but joined-up. Managed migration was developed later as a 'narrative rather than strategy, because the background was already there…This thing had grown up over many years and there was never I suppose a real strategy—top to bottom strategy—for managed migration' (Interview with former CV, 2012). This demonstrates that successive rounds of 'normal policy change' can highlight inconsistencies with the previous policy paradigm to the new set of circumstances, and in turn generate an atypical or third-order change.

6.4.6 Policy Entrepreneurs

While institutional changes and departmental interests were critical to this narrative of policy change, initiatives, policies and proposals, however innovative, need political interests and therefore actors supporting them to materialise. The historical institutionalist argues that ideas are not free-floating abstractions but rather require interests and actors to legitimise and mobilise them—the more powerful the actor the better. In the case of immigration policy change, one minister in particular could be regarded as an agenda setter or a policy entrepreneur—Home Secretary from 2001 until 2004, David Blunkett.

A cabinet reshuffle in 2001 saw David Blunkett—who had been Education and Employment Secretary—being promoted to Home Secretary. Upon this appointment there 'was literally a big change within a week' (Interview with CV, 2012), further demonstrating the relatively dormant period of immigration policymaking under previous Home Secretary Jack Straw. According to Blunkett his experience at DfEE and his knowledge of the work permit system influenced his assessment of immigration policy: 'It [DfEE] made me liberal with a small l. It made me see that immigration was a multifaceted issue' (Interview with David Blunkett, 2012). Administratively Blunkett brought the work permit scheme with him to the Home Office from DfEE as he 'couldn't see how you could logically not put them together' (Interview with David Blunkett, 2012). According to a special advisor, Blunkett came to the Home Office with a more positive and strategic approach than previous Home Secretaries and brought wider labour market thinking. Blunkett was purportedly very proud of the work permit system under DfEE as he had 'revolutionalised' the system, getting the time to obtain a work permit down to days rather than months, which 'won him brownie points from the business sector' (Interview with Don Flynn, 2011). A former civil servant explains how Blunkett's appointment was so significant to policy change:

> David Blunkett had a very strong interest in the subject and the economic value immigrants can bring to our society; he'd said those things when he was in education and he had run work permits which was actually a more open scheme...the implementation was run by DfEE and that was good, they were really good in that department, they really got it. So the Treasury had been working with the Home Office for some time to try and get them to do this thing, and they had got somewhere but not terribly far. And then there was a moment when Straw left in 2001 and Blunkett arrived. That was a really big change in the Home Office, literally when Blunkett came in...It was just that with Blunkett there it meant that they could go for a much more ambitious criterion. (Interview with former civil servant, 2012)

Under Blunkett's leadership the number of work permits granted increased, attracting international students became part of the managed migration agenda, and the overall rhetoric of the Home Office changed from control to 'managed migration'. Having experienced the work permit system in DfEE David Blunkett's personal interest in the area saw material changes to immigration policy and a shift in rhetoric, going so far as to suggest that there was 'no obvious limit' to immigration on a BBC *Newsnight* broadcast (Blunkett quoted in BBC 2003).

Blunkett's appointment to Home Secretary in 2001 was a catalyst for institutional and, more pertinently here, policy change. The ideas and interests for such reforms were already circulating between the Treasury and DfEE, but the institutional norms and embedded policy framing in the Home Office were somewhat resistant to this new policy framing. Not that the Treasury needed to persuade Blunkett: 'David didn't need pushing and he wouldn't have been pushed by Treasury officials' (Interview with former SpAd, 2014). Blunkett sought to, and was appointed to, 'shake things up' in the Home Office (Interview with former Official,

2011), and a powerful actor was needed to overcome the institutional inertia or bias towards restriction that permeated Home Office policymaking. Blunkett was particularly frustrated with Home Office culture as even the briefest scan of his memoir reveals, and he readily conceded that he came in as 'a steamroller over the way the department worked and its wider policies' (Blunkett 2006, 285). He and Blair wanted a 'complete revamp' of the department administratively and culturally (Ibid, 298).

However, this is not to place undue emphasis on one actor, nor is this to suggest that Blunkett was the driver for the overarching shift in policy objectives. It is rather to acknowledge that ideas need interests, that ministers are capable (if they have the political will) of overhauling policy paradigms, but most importantly it demonstrates the contingency between timing, interests and the need for powerful actors to mobilise ideas to dislodge institutionalised policy framings. In other words, ministers can matter but only in the right administrative context combined with other favourable dynamics.

6.5 Evidence, Knowledge and Policy Transfer

'It's so political, evidence-based policy' (Interview with civil servant, 2012)

Given that research and evidence were purportedly an integral part of the 'policy wheel' under the Labour governments, we now turn to whether the commissioning of evidence, and the production of new knowledge, under the rubric of evidence-based policy proved to be important in shifting the policy objectives.

The accumulation of evidence and knowledge is a process of 'policy learning', which includes evidence from government, think tanks and academia. But the process of policy learning is not confined to the production of tangible evidence. Learning includes taking heed from past policies including failures (Hall 1993), drawing on lessons from other policy areas, and borrowing ideas from other countries otherwise known as 'policy transfer' (Rose 1991). Historical institutionalists contend that new knowledge or learning can, if only occasionally, alter policy frames and in turn change policy (Sabatier and Jenkins-Smith 1993, 104; Hall 1993).

How evidence and knowledge are used in policymaking has long been disputed amongst practitioners, commentators and scholars. In contrast to the instrumentalist theories that dominated the field in the 1980s, Majone (1989) argued the use of evidence in policymaking is bound up in the skills of persuasion and argumentation. According to Majone, evidence is selected information 'from the available stock [which] is introduced at a specific point in an argument to persuade the mind that a given factual proposition is true or false' (Majone 1989, 48). Majone argued that evidence is politically constructed and driven by political interests.

Addressing the uses and functions of expert knowledge in immigration policymaking specifically, Boswell (2009) claims that evidence can perform (1) an instrumental function, (2) a legitimising function or (3) a substantiating function (although these are not mutually exclusive). Knowledge is used instrumentally when governments use research to fill gaps in their knowledge, 'in order to adjust policy in a way that will achieve the desired societal impacts' (Boswell 2009, 4). In contrast, expert knowledge can work to fulfil political functions that are not necessarily instrumental. The first of these symbolic uses of evidence is what Boswell calls a legitimising function, where 'by being seen to draw on expert knowledge, an organisation can enhance its legitimacy and potentially bolster its claim to resources or jurisdiction over particular policy areas' (Boswell 2009, 7). The final substantiating function expresses the way in which expert knowledge can help 'substantiate an organisation or political party's preferences, and undermine those of rival agencies or organised interests' (Boswell 2009, 7). Like Majone, Boswell emphasises the political function of knowledge in policymaking, and contends that 'political organisations are likely to draw on knowledge as a source of legitimation rather than in order to improve output' (Boswell 2009, 13).

In the 1990s research on migration within government was sparse. There was no migration research unit, no economists working on migration and beyond producing annual statistics there was no demand for producing any type of research, at least from central government. But with Labour's pursuit of evidence-based policy, and in turn an increase in funding to research crime and policing, resources were eventually poured into migration research (Interview with CV, 2012). The most significant development was the establishment of the Immigration Research and Statistics Service (IRSS) in 2000, which amongst other developments set up a new migration research unit. The new research agenda on immigration was officially launched a year later in 2001 at a major conference organised by the Home Office (2001) entitled 'Bridging the Information Gaps'.

Over Labour's time in government, innumerable research reports on migration were produced. Issues concerning the integration of immigrants and refugees dominated much of the research commissioned. However, the majority of this research cannot be said to have impacted on policy in any discernible way, in part because the notion of evidence-based policy was 'only patchily internalised in Home Office thinking...it seems to represent a case of coercive isomorphism whereby the organisation felt obliged to adopt the trappings of a research function' (Boswell 2009, 157). Findings amongst interviewees suggested that the research agenda developed in isolation from the policy agenda, in the sense that there was a disconnect between the policy direction, focus and objectives, and the research being produced. One civil servant recalled that at the time there was 'evidence based policy fatigue', as there were 'too many interventions and too many ideas' (Interview with CV, 2012). Another interviewee commented that the 'EBP bug' became redundant to the extent that research was constantly being produced but the research had 'no relevance to policy' (Interview with CV, 2012). This was especially acute because the output was focused on producing reports that 'become almost irrelevant after the years it takes to publish' rather than the utilisation of evidence for strategic policymaking (Interview with CV, 2012).

In terms of the types of evidence deemed instrumental in immigration policymaking under the Labour administrations, there was a clear preference for quantitative research over more qualitative social research. This is in part because as one interviewee commented 'it is extremely hard to get good quality evidence on the social impacts side of things and to find a sensible way to interpret it'. For example, many reports were concerned with the drivers for migration, the results of which were difficult to operationalise into policy.

Conversely, it seems for research and knowledge to be utilised by government in policymaking requires the research to be policy relevant in the first instance, therefore somewhat contradicting the notion of evidence based policymaking and rather implying 'policy based evidence' (Interview with civil servant, 2012).

The very scope and definition of evidence and knowledge in terms of policy impact also evolved under the Labour governments. In turn, the remit of policy analysts in the migration field broadened; impact assessments, economic forecasts, evaluating and monitoring policy impacts all figured as part of the evidence base of immigration policy. With the wider reforms in policymaking and policy delivery—New Public Management— evaluative research and opinions from stakeholders were highly valued. For example, as discussed in Chap. 4 the PBS and the MAC were designed according to consultation feedback from various stakeholders. However,

Ann Singleton, who was involved in a government review of evidence use in immigration policymaking, commented that although there were extensive consultations on how to implement the PBS, the 'associated research appeared to be limited to how best the new system should be implemented' (Interview with Ann Singleton, 2013). The PBS was foremost a 'political decision' (Interview with Ann Singleton, 2013).

It is no secret that there was an element of policy transfer involved in the designing of the PBS, as Immigration Minister Liam Bryne acknowledged when launching the system 'we learnt from the most effective systems in the world, in this case Australia' (Byrne 2008b). Home Office teams consulted (and seconded) with the Australian and Canadian governments to discover and learn what types of evidence they used to underpin and demarcate their PBSs (Interview with CV, 2012). The Home Office particularly 'borrowed' ideas on sponsorship roles and functions directly from the Australian and Canadian systems (Interview with CV, 2012). One Home Office civil servant commented that they were 'more likely to have links and meetings with the Australians then other UK departments' when developing the PBS (Interview with CV, 2012). To the extent that these sources constitute evidence and knowledge, it is fair to say that knowledge utilisation shaped the operationalisation and functionality of the PBS, and therefore evidence in this sense shaped what Hall (1993) calls a 'second order change'.

Yet whether research, evidence and knowledge played a role in prompting the 'third order change' (Hall 1993) under study here is far more ambiguous. Proving that evidence has a direct impact on policy outcomes is challenging. Nonetheless, as we will see, on the whole, evidence played an important but mostly substantiating role in policy change (Boswell 2009). The government were keen to cultivate evidence of the positive economic impact of immigration and subsequently commissioned extensive research into this theme. But such evidence served mainly to depoliticise and frame the issue as one of economic rationality, indeed economic necessity, thus bolstering the case for expansion but not precipitating it.

As part of the JUG strategy and the evidence-based policymaking ideal, the Labour administration set up two units—the PIU and the Prime Minister's Forward Strategy Unit—both based in the Prime Minister's Office. These were later merged in 2002 to form the Strategy Unit. These were established to support government departments in establishing effective strategies and policies and provide policy advice directly to the Prime Minister with an emphasis on an evidence-based approach. Precisely because immigration is a cross-departmental issue the PIU, and later the Strategy Unit, conducted in-depth analyses of immigration policy. This resulted in a strategic review of immigration and a report in 2001, the latter of which according to one of the authors 'ignited a debate within Whitehall' on the framing of immigration as an economic issue (Interview with Jonathan Portes, 2011).

The PIU published the first thorough economic analysis of immigration flows in 2001, *Migration: an economic and social analysis* (Glover et al. 2001). The paper was particularly significant because it was the first cross-departmental research report on immigration, receiving input from the Home Office, Treasury, DTI, Health and DfEE. The report was thus a product of the evidence-based policy model and the JUG strategy. Particularly significant was that the report was led by a former Treasury and Cabinet economist, Jonathan Portes, who 'combined with the Treasury had a very strong sense that migration was good for growth' (Interview with Senior Civil Servant, 2012). Additionally, non-government experts such as Sarah Spencer from the IPPR were seconded in for the report (a further manifestation of JUG and EBP). The 2001 report was made because of earlier investigations in the PIU, as Jonathan Portes recalled:

> It came because of an earlier report I did in the PIU in 1999 on strategic challenges. There was a theme that came through very strongly– that immigration was something that was quite important part of what was going on in the UK economy and that the government hadn't been thinking about it at all and that maybe they ought to be thinking about what the evidence base said and do some analysis. (Interview with Jonathan Portes, 2011)

The report provided an evidence base and thus legitimated the expansive reforms, as Portes suggests:

> It certainly changed the Whitehall dynamic, there was a sudden realization that immigration was an interesting issue, that there was quite a lot of economics, that it was an economic issue and it hadn't been thought of as an economic issue before.

In his insightful account of the use of ideas and evidence in immigration policy, Balch (2010) argues that evidence generally and in particular the 2001 PIU report played a crucial role in shifting the policy frame. He places emphasis on the role of experts such as Spencer and

Portes and the knowledge they imported and argues that its 'effects become key to the push for policy change' (Balch 2010, 154). However, the Treasury was pushing immigration policy in this direction prior to the report but the Treasury did not have the resources to execute the research. While according to Portes, the impetus for the research was not driven by policy initiatives, most interviewees surmised that the Labour government intended to embrace expanding immigration flows before the report was produced and that the government chose the policy before it really began to think about the evidence. Even Portes claimed that the government 'chose by default a broadly accommodative policy' (Interview with Jonathan Portes, 2012). Other interviewees similarly commented that while the PIU report was important for 'marshalling arguments' about the economic benefits of immigration, and shifting how Whitehall as a whole thought about the issue, the report itself did not influence the policy or Blunkett personally as Home Secretary (Interview with former SpAd, 2014). The 2001 report was an important part of 'marshalling technocratic arguments' (Boswell 2009, 128), but such research did not play an instrumental function, nor was it a necessary component of the policy shift.

The PIU report was followed by a second, more comprehensive report by the Strategy Unit (which was never made publicly available), which was attempting to achieve a JUG approach to immigration by consulting extensively across Whitehall:

> One of the things we were trying to do—and I think we largely failed—was to achieve coherence. Migration policy had been made in a fairly ad hoc way by different parts of government. The Home Office had one set of perspectives, the DTI had a completely different set of perspectives, and education was yet another one. We were trying to look in a more rounded way at the different trade-offs and issues and the long term dynamics but I can't claim we succeeded in that or rather I think we succeeded to a degree analytically but not in terms of getting Cabinet to really stand back. (Interview with former director of Strategy Unit, 2012)

Whilst the report did not succeed in achieving a cohesive or multidimensional policy, its recommendations for a PBS were taken up, although Blair had already decided he wanted a PBS (Interview with SpAd, 2011). The Glover et al. report however provided a public evidence base for legitimising the policy shift, but this was only significant because of the rubric of evidence-based policy, which gave the report an added credibility. Research had previously been commissioned into the economic impacts of immigration in the post-war period, and the results (whilst encompassing the usual methodological difficulties) found that immigration was on the whole economically beneficial. These reports were dismissed by previous governments, but because the Labour government were so persuaded by the modus operandi of evidence-based policymaking, at least for legitimising purposes, the 2001 report carried more weight.

While evidence did not play an instrumental function, the pursuit of EBP did provide the government with leverage for the managed migration agenda, in particular the much quoted figures of immigrants boosting GDP, and the Glover et al. (2001) report. With the advent of evidence-based policy the terms of the debate shifted,

> ...from traditional forms of political debate revolving around interests and bargaining, towards criteria based on economic and technological development goals...the new discourse meant debating the pros and cons of immigration within a new paradigm, one based on economic rationality. (Boswell 2009, 143)

The steering towards rationalising and depoliticising policy (Burnham 2001) allowed the government to present a technocratic basis for expanding immigration policy. The evidence produced bolstered the case for managed migration, reinforcing the arguments for liberalisation and providing the government added legitimacy in conjunction with support from the business community. Nonetheless, to the extent that evidence did play a role in economic immigration policymaking in this period, it served as a substantiating function, providing 'sound-bites' to support the new agenda (Boswell 2009, 158).

6.6 Conclusion

This chapter has shown that the overarching policy change, or third-order change (Hall 1993), was not centrally steered, but was rather a by-product of a series of policy reforms, driven by departmental agendas, which culminated in an expansive immigration regime. Each policy reform had its own drivers associated with specific departments. The managed migration rhetoric employed in 2002 was a way of providing a coherent narrative for policy changes that had already transpired. This demonstrates the importance of taking a disaggregated view of the state, assuming a monolithic lens of the state would miss crucial decision-making processes that were so pertinent to this case of policy change.

The Home Office prior to 1997 had dominated immigration policymaking. Because of the issues within the Home Office remit, the Department had developed a cautious and socially conservative culture. Such culture meant that immigration was framed as an enforcement issue similar to crime, with the overriding objective of immigration to be controlled and minimised. This culture persisted in the framing of asylum policy and is perhaps prevalent once again in the 2010s political climate. However, the Home Office monopoly on economic immigration was fractured under the Labour governments, and with it a new framing was embedded.

The impetus for the policy change came from the Treasury, a Department that was especially powerful under Gordon Brown. Other departments such as DfEE similarly launched at the time non-contentious schemes to boost departmental agendas. The Treasury, along with other business-orientated departments, imported their culture and in turn economic framing to immigration policy for the first time, symptomatic of an endogenous process of institutional conversion (Thelen 2004). As a result the policy frame was changed from a control focus to an economic policy.

Yet ideas do not materialise by themselves. As innovative as an idea may be, without agenda setters pushing for a change in a policy frame, the idea can be overlooked. In other words without the support of powerful political elites, namely ministers, a new policy frame will not be embedded. Institutions are after all resilient to change but the appointment of a new minister can, if there is the political will, work as a catalyst to overcome institutional inertia. Ideas and interests are thus interdependent. In this case, David Blunkett in particular played an important role as a policy entrepreneur, encouraging immigration to be seen as an economic issue.

It was however Labour's third-way ideology (see Chap. 5 for discussion) and the way this was reflected and embedded in the institutional settings of policymaking that explain how these departments became involved in what hitherto had been a closed policy network. With the idea that policy should be based on 'what works', Labour attempted to change policymaking practices with a joined-up government strategy and evidence-based policymaking. The institutional changes to policymaking, which brought new actors into the circle, were politically driven changes, influenced by the broader third-way ideology of Labour. The evidence commissioned into migration as a result of the evidence-based policy rubric did not trigger the policy reforms. But the evidence nonetheless aided the government in marshalling rationalistic arguments for expanding immigration. While these initiatives were not wholly successful, the impact of rationalising policy based on evidence and extending the 'turf' of immigration policy to other departments as well as widening the policy network through secondments and advisers from outside government had a clear impact on the policy frame. Paradoxically, whilst turf was extended to non-Home Office departments, the result was an immigration policy that was anything but joined up. It was only later, after reforms had been passed, that the Home Office attempted to consolidate such initiatives under the concept of managed migration.

Institutions, far from being 'frozen residual structures' (Thelen 2004, 8) as many new institutionalists implicitly assume, are the 'objects of ongoing political contestation, and changes in the political coalitions on which institutions rest are what drives changes in the form institutions take and the functions they perform in politics and society' (Thelen 2004, 31). Thus institutions, whilst important structures that condition policy framings and indeed policy implementation, are essentially political constructions, which, although resistant to change, are adaptive and susceptible to political influence from the governing party. We now turn to the concluding chapter, which summarises the findings and draws the threads together to provide an overarching explanation of policy change.

REFERENCES

Balch, A. (2010). *Managing Labour Migration in Europe: Ideas, Knowledge and Policy Change*. Manchester: Manchester University Press.

Bauer, T., & Zimmerman, K. (1999). *Assessment of Possible Migration Pressure and Its Labour Market Impact Following EU Enlargement to Central and Eastern Europe* (IZA Research Report No. 3). London: Department for Education and Employment.

BBC. (2003, 13 November). Blunkett: No Limit on Migration. *BBC*. Available from: http://news.bbc.co.uk/1/hi/uk_politics/3265219.stm. Accessed 16 Jan 2010.

Béland, D. (2005). Ideas and Social Policy: An Institutionalist Perspective. *Social Policy and Administrative, 39*(1), 1–18.

Berman, S. (1998). *The Social Democratic Moment: Ideas and Politics in the Making of Interwar Europe*. Cambridge: Harvard University Press.

Blair, T. (1997, December 8). Bringing Britain Together. *Speech*. London. Available from: http://www.britishpoliticalspeech.org/speech-archive.htm?speech=320. Accessed 11 July 2011.
Blair, T. (2004, April 27). *Speech to Confederation of British Industry on Migration*. London. Available from: http://www.theguardian.com/politics/2004/apr/27/immigrationpolicy.speeches. Accessed 21 Mar 2010.
Blair, T. (2010). *Tony Blair: A Journey*. London: Hutchinson.
Blunkett, D. (2000, February 2). *Influence or Irrelevance: Can Social Science Improve Government*. Speech to the Economic and Social Research Council. Available from: www.bera.ac.uk/beradev2002/root/archive/ri/no71/index.html. Accessed 18 Feb 2010.
Blunkett, D. (2002, June 26). *Home Secretary's Speech to Social Market Foundation*. London: Social Market Foundation
Blunkett, D. (2006). *The Blunkett Tapes: My Life in the Bear Pit*. London: Bloomsbury.
Blunkett, D., & Richards, D. (2011). Labour In and Out of Government: Political Ideas, Political Practice and the British Political Tradition. *Political Studies Review, 9*(2), 178–192.
Bogdanor, V. (2005). Introduction. In V. Bogdanor (Ed.), *Joined-Up Government* (pp. 1–19). Oxford: Oxford University Press.
Boswell, C. (2009). *The Political Uses of Expert Knowledge: Immigration Policy and Social Research*. Cambridge: Cambridge University Press.
Burch, M., & Holliday, I. (2004). The Blair Government and the Core Executive. *Government and Opposition, 39*(1), 1–21.
Burnham, P. (2001). New Labour and the Politics of Depoliticisation. *The British Journal of Politics and International Relations, 3*(2), 127–149.
Byrne, L. (2008a). *Home Affairs Select Committee Hearing with Liam Byrne and Lin Homer on November 27*. Available from: http://www.publications.parliament.uk/pa/cm200708/cmselect/cmhaff/123/7112701.htm. Accessed 10 July 2011.
Byrne, L. (2008b, February 6). *The Case for a New Migration System*. Speech to the Local Government Association. Available from: http://press.homeoffice.gov.uk/Speeches/sp-lb-lga-feb-08. Accessed 10 July 2011.
Cabinet Office. (1999a). *Modernising Government*. London: Cabinet Office.
Cabinet Office. (1999b). *Professional Policymaking in the 21st Century*. London: Cabinet Office.
Cabinet Office. (2000). *Wiring It Up: Whitehall's Management of Cross-Cutting Policies and Services*. London: Performance and Innovation Unit.
Cabinet Office. (2001). *Modernising Policy Development*. London: Cabinet Office.
Cabinet Office. (2006). *Capability Review of the Home Office*. London: Cabinet Office.
Callaghan, J. (1983). Cumber and Variableness. In *The Home Office: Perspectives on Policy and Administration* (pp. 19–22). London: Royal Institute of Public Administrations.

Caulkin, S. (2006). Why Things Fell Apart for Joined-up Thinking. *The Observer* [Online]. Available from: https://www.theguardian.com/society/2006/feb/26/publicservices.politics

Chapman, R. A. (2002). *Treasury in Public Policy-Making*. London: Routledge.

Clark, T. (2002). New Labour's Big Idea: Joined-up Government. *Social Policy and Society, 1*(2), 107–117.

Council of Europe. (2000). *Europe's Population and Labour Market Beyond 2000*. Germany: Council of Europe.

DTI (Department for Trade and Industry). (1998). *Our Competitive Future: Building the Knowledge Driven Economy*. London: DTI.

Dustmann, C., Casanova, M., Fertig, M., Preston, I., & Schmidt, C. (2005). *The Impact of EU Enlargement on Migration Flows Home Office Report 25/03*. London: Home Office.

Düvell, F., & Jordan, B. (2003). Immigration Control and the Management of Economic Migration in the United Kingdom: Organisational Culture, Implementation, Enforcement and Identity Processes in Public Services. *Journal of Ethnic and Migration Studies, 29*(2), 299–336.

Fawcett, P., & Rhodes, R. A. W. (2007). Central Government. In A. Seldon (Ed.), *Blair's Britain 1994–2007*. Cambridge: Cambridge University Press.

FCO (Foreign and Commonwealth Office). (2002, December 10). *Jack Straw Announces the Extension of the Free Movement of People Rights to EU Candidate Countries on Accession*. FCO Press Release.

Finlayson, A. (2003). *Making Sense of New Labour*. London: Lawrence & Wishart.

Gilpin, N., Henty, M., Lemos, S., Portes, J., & Bullen, C. (2006). *The Impact of Free Movement of Workers from Central and Eastern Europe on the UK Labour Market*. London: DWP.

Glover, S., Gott, C., Loizillon, A., Portes, J., Price, R., Spencer, S., Srinivasan, V., & Willis, C. (2001). *Migration: An Economic and Social Analysis*. London: Home Office.

Gott, C., & Johnson, K. (2002). *The Migrant Population in the UK: Fiscal Effects, RDS 77*. London: Home Office.

Government Actuary Department. (2002). *National Population Projections, 2000 Based, Series PP2, No. 23*. London: The Stationery Office.

Hall, P. (1993). Policy Paradigms, Social Learning, and the State: The Case of Economic Policymaking in Britain. *Comparative Politics, 25*(3), 275–296.

Hansard HC Deb. vol. 391 cols. 534-6, 28 October 2002.

Hansard HC Deb. vol. 418 cols. 870-8, 2 March 2004.

Hansen, R. (2000). *Citizenship and Immigration in Post-war Britain*. Oxford: Oxford University Press.

Head, B. W. (2010). Reconsidering Evidence-Based Policy: Key Issues and Challenges. *Policy and Society, 29*(2), 77–94.

Heffernan, R. (2006). The Prime Minister and the News Media: Political Communication as a Leadership Resource. *Parliamentary Affairs, 59*(4), 582–598.
Hennessy, P. (2002). The Blair Government in Historical Perspective: An Analysis of the Power Relationships Within New Labour. *History Today, 52*(1), 21–23.
HM Government. (2009). *PSA Delivery Agreement 3: Ensure Controlled, Fair Migration that Protects the Public and Contributes to Economic Growth*. London: Home Office.
HM Treasury. (2000a). *Budget 2000*, March. London: Treasury.
HM Treasury. (2000b). *Productivity in the UK: The Evidence and the Government's Approach*. London: Treasury. Available from: www.hm-treasury.gov.uk/d/ACF1FBA.pdf. Accessed 10 Mar 2013.
HM Treasury. (2000c). *Spending Review 2000: New Public Spending Plans 2001–2004*. London: HM Treasury.
HM Treasury. (2002). *Trend Growth: Recent Developments and Prospects*. London: HM Treasury.
HM Treasury. (2003). *Budget Report 2003: Building a Britain of Economic Strength and Social Justice*. London: HM Treasury. Available from: www.hm-treasury.gov.uk/d/budget_2003.pdf. Accessed 27 Jan 2013.
HM Treasury. (2004). *Long-Term Global Economic Challenges and Opportunities for the UK*. London: HM Treasury. Available from: www.hm-treasury.gov.uk/d/pbr04global_421.pdf. Accessed 24 Apr 2013.
Home Office. (2001). *Bridging the Information Gaps: A Conference of Research on Asylum and Immigration in the UK*. London: Home Office.
Home Office. (2002). *Secure Borders, Safe Havens*. London: Home Office.
Home Office. (2007). *Departmental Report, 2007*. London: Home Office.
House of Lords. (2008). *The Economic Impact of Immigration*. House of Lords Select Committee on Economic Affairs, 1st Report of session 2007–8. London: Stationery Office.
Howlett, M., Ramesh, M., & Perl, A. (2009). *Studying Public Policy: Policy Cycles and Policysubsystems*. Oxford: Oxford University Press.
Kavanagh, D., & Richards, D. (2001). Departmentalism and Joined-up Government: Back to the Future? *Parliamentary Affairs, 54*(1), 1–18.
Liaison Committee. (2001/2002). *Minutes of Evidence*. London: UK Parliament. Available from: http://www.publications.parliament.uk/pa/cm200102/cmselect/cmliaisn/cmliaisn.htm. Accessed 27 Dec 2012.
Ling, T. (2002). Delivering Joined-up Government in the UK: Dimensions, Issues and Problems. *Public Administration, 80*(4), 615–642.
Majone, G. (1989). *Evidence, Argument, and Persuasion in the Policy Process*. New Haven: Yale University Press.
Mandelson, P. (2010). *The Third Man*. London: Harper Press.
Mandleson, P. (1996, February 25). Interview with Peter Mandleson. *BBC*.

Moseley, A. (2009, September). *Joined-up Government: Rational Administration or Bureaucratic Politics?* Paper Presented to the Politics of Public Services Panel, Public Administration Committee Annual Conference.

Mulgan, G. (2005). Joined-up Government: Past, Present, Future. In V. Bogdanor (Ed.), *Joined-up Government* (pp. 175–187). Oxford: Oxford University Press.

NAO (National Audit Office). (2001). *Modern Policy-Making: Ensuring Policies Deliver Value for Money. Report by the Controller and Auditor General, HC 289 Session 2001–2002: November 2001.* London: National Audit Office.

OECD. (2001). *Trends in Immigration and Economic Consequences.* London: OECD.

OECD. (2009). *Sopemi Country Notes.* Available from: http://www.oecd.org/dataoecd/43/0/44068261.pdf. Accessed 1 Aug 2010.

Page, E. C. (2005). Joined-up Government and the Civil Service. In V. Bogdanor (Ed.), *Joined-Up Government* (pp. 139–156). Oxford: Oxford University Press.

Pollitt, C. (2003). Joined-up Government: A Survey. *Political Studies, 1*(1), 34–49.

Richards, D., & Smith, M. (1997). How Departments Change: Windows of Opportunity and Critical Junctures in Three Departments. *Public Policy and Administration, 12*(2), 62–79.

Richards, D., & Smith, M. J. (2005, April). *Institutional Reform for Political Control: Analysing the British Labour Government's Approach to the Pathologies of Governance.* Paper Presented to a Scanor Workshop in Collobaration with SOG.

Richards, D., & Smith, M. (2007). Central Control and Policy Implementation in the UK: A Case Study of the Prime Minister's Delivery Unit. *Journal of Comparative Policy Analysis: Research and Practice, 8*(4), 325–345.

Rose, R. (1991). What Is Lesson-Drawing? *Journal of Public Policy, 11*(1), 3–30.

Sabatier, P. A., and Jenkins-Smith, H. (1993). *Policy Change and Learning: An Advocacy Coalition Approach.* Boulder: Westview Press.

Schifferes, S. (2002, June 17). Analysis: Who Gains from Immigration? *BBC News.* Available from: http://news.bbc.co.uk/1/hi/business/2019385.stm. Accessed 17 Jan 2012.

Shaw, C. (2001). *United Kingdom Population Trends in the 21st Century.* Population Trends 103. London: The Stationery Office.

Smith, J. (2007, December 5). *Shared Protections, Shared Values: Next Steps on Migration.* Speech to the London School of Economics and Political Science, London. Available from: http://www.lse.ac.uk/government/research/resgroups/MSU/event-jacquismith.aspx. Accessed 17 June 2010.

Spencer, S. (2011). *The Migration Debate.* Bristol: Policy Press.

Thelen, K. (2004). *How Institutions Evolve.* Cambridge: Cambridge University Press.

TNA FCO 50//532 Immigration Policy Review of FCO Functions; Letter to Mr Hawley from H. E. Rigney, Inspection Report on Migration and Visa Departments, August 1975.

TNA FCO 50/585 Meeting on Immigration Between FCO and Home Office Ministers; Memo to Secretary of State from Mr Luard, 24 May 1976.

Wells, P. (2007). New Labour and Evidence Based Policymaking: 1997–2007. *People, Place & Policy Online*, 1(1), 22–29.

Williams, P. (2004). Who's Making UK Foreign Policy? *International Affairs*, 80(5), 911–929.

Wright, C. (2010). *Policy Legacies and the Politics of Labour Immigration Selection and Control* (PhD Dissertation). Cambridge University.

CHAPTER 7

An Unintended Consequence...

7.1 Introduction

Twenty-five years ago Britain's Prime Minister described large-scale immigration into Europe as one of the 'most striking failures of our age' (Thatcher 1992). Just over ten years later, the Home Secretary was claiming that there 'is no obvious limit to immigration' (Blunkett 2003). A profound shift transpired in immigration policy between 1997 and 2010 from a reluctant country of immigration to a retrospectively expansive regime. This policy transformation is perplexing and interesting in its own right, but more than just an isolated anomaly in British politics, the changes made have had a lasting impact on both the population of Britain—around half (3.8 million) of all usual residents of England and Wales on census day who were born outside the UK arrived in the UK between 2001–2011 (ONS 2012)—and the current political debate on immigration. Indeed this period is and was the making of a migration state.

This book has explored the development of economic immigration policy in Britain under the Labour administrations, setting out to answer the question of how and why it changed between 1997 and 2010, from a restrictive approach to a comparatively expansive policy. Drawing on three existing approaches, the book has examined how organised interests, party politics and institutions shaped immigration policy and triggered policy change. This concluding chapter summarises the key findings in answer to the research questions, presents an overarching explanation for why policy changed and offers explanatory insights for scholars of policy change generally.

7.2 Summary and Key Findings

The key arguments of the book are fourfold. First, the evidence has shown that there was no single cause of policy change. The shift in direction was rather a result of a combined set of favourable conditions. Each approach—organised interests, party politics and historical institutionalism—offers a partial explanation but no one approach can sufficiently explain why such a radical policy shift occurred. Second, the shift to managed migration was a consequence of an accumulation of policy reforms from different departments with different agendas. The overarching change in the policy framework was not preconceived and the repercussion was not intended. Third, the logic of the policy reforms was informed by the Labour Party's third-way ideology, in particular their fixation with globalisation and friendly capitalism. Fourth, immigration policy remained elite-driven in this period.

7.2.1 Organised Interests

Organised interests did not drive policy in a liberalising direction. Non-state actors, while broadly supportive of the reforms, did not lobby government for such a radical change. The interest groups interviewed remarked that their efforts to lobby government on economic immigration began after the policy reforms had been enacted. Indeed many interviewees described their lobbying efforts as 'reactive' to government policy. There was no evidence to suggest that interest groups were lobbying government in any substantial way in the late 1990s or early 2000s. This is illustrated by the limited interest, resources and engagement that interest groups invested in the issue. Indeed many policy decisions went ahead despite lobbying from interest groups.

There is some evidence to suggest that employers were disgruntled at the various administrative hurdles required to obtain work permits in the late 1990s. However, it would be an overstatement to suggest that this was the impetus for a change in government thinking. Employers' lobbying efforts were confined to reducing visa costs and red tape in the 1990s, and these efforts were minimal at best.

7.2 SUMMARY AND KEY FINDINGS 203

When talking about the role of non-state actors, it is impossible to ignore key figures from the centre-left think tank IPPR. Arguably actors from IPPR represented policy entrepreneurs (Kingdon 1995) in this story; the organisation persistently spoke of the positive economic benefits of immigration throughout the 1990s and early 2000s (Spencer 1994). Under the Major government this advocacy had little impact, but it was a perspective that resonated with New Labour's governing philosophy. Described as 'Labour's civil service' (Taylor quoted in The Observer 2003), the close links between the Labour Party and IPPR were evident, and particular actors such as Blunkett's special advisor Nick Pearce (formerly IPPR) have been widely acknowledged as having a significant influence on these reforms (Spencer 2007). A handful of other actors could be mentioned here, but the point is not to exaggerate the role of individuals. Rather, IPPR as an organisation had a prominent role in shaping, developing and reinforcing the ideas of the Labour government, often doing so by producing evidence to substantiate their beliefs. Alex Balch has stressed this point and suggests that 'it is possible that elements of the Labour government had at least partially assimilated the recommendations outlined by Spencer in 1994' (Balch 2010, 148). It is hard to discern and measure the direct impact the organisation had on policy. Nonetheless the element of congruence between the ideas and arguments of IPPR at this time and Labour's immigration policy is evident. However, their role should not be overstated; the government appeared to be independently thinking in these terms already.

While organised interests were not responsible for the expansionary policies, the policy community were advantageously broadly supportive of the reforms and employers and employer associations certainly favoured an expansive immigration policy. Fortunately this was seen as consistent with the government's objectives. Interest group support was, however, rhetorically used by political elites to legitimise policy decisions. One former Home Secretary suggested that the government wanted employers to be more vocal and campaign publicly in support of the changes to help government 'carry the day' (Interview with David Blunkett, 2012).

While interest groups did not cause the shift, their power to influence policy intensified during the Labour administrations. However, their gaining influence was a result of the Labour governments being keen to engage with stakeholders rather than pressure from interest groups. Increased stakeholder engagement was therefore a consequence, not a cause, of policy change. While the overarching policy shift was not a product of interest groups lobbying government, the contributions from interest groups on the PBS were imperative. The notion of having a PBS was already underway, but the operationalisation and specifics of the system were based on interest group contributions via consultations; thus they shaped 'second order' changes (Hall 1993) to policy. Whilst there was limited evidence to suggest that organised interests were responsible, their role in making and shaping policy was enhanced under the Labour administrations, and their support and input have been critical for policymaking from the mid 2000s onwards.

The organised interests approach, and specifically Gary Freeman's client politics model, cannot explain why the Labour government liberalised policies. Conversely, the state and the actors that comprise it proved to be the crucial actors in this narrative, confirming Statham and Geddes' (2006) finding that immigration policymaking, in the UK context, is elite-driven. The incompatibility between Freeman's model and the British case is perhaps because the model is based on the US federal system, which offers multiple avenues for interest groups to lobby government. This suggests that Freeman's predictions might have greater authority in a political system perceived to be more open to the interests of big business than in a parliamentary system such as the UK where constituency MPs are more receptive to the (usually) anti-immigration preferences of their voters (Balch 2010, 21). Moreover, Congress plays a more significant role in shaping policy in the US. In comparison, the executive leads on policy in a fairly autonomous manner in the UK (Sassen 1999, 187). It seems in this case the analytic model does not travel across the Atlantic particularly well.

7.2.2 Party Politics

The book set out to assess whether party politics—namely party competition and party ideology—contributed to the shift in immigration policy. Chapter 5 examined whether policy change was a consequence of a change in administration, party ideology and/or party strategy. The empirical findings showed that the change to the party composition of government was imperative. If we consider the counterfactual possibility of the Conservative Party remaining in power in 1997, it is highly unlikely that the reforms would have been made, with the possible exception of the A8 decision. However, centre-left parties do not unequivocally favour expansive immigration policies. Comparisons across Western Europe (see Koopmans et al. 2012, 1229) as well as Labour's history on immigration demonstrate that this is not the case. Ideologically there is no fixed or given position on immigration for the centre left; indeed immigration brings the ideological tensions of social democratic parties to the fore, between on the one hand international solidarity of the vulnerable working class and on the other welfare state/labour market protectionism (Odmalm 2011, 1071). Nonetheless, being part of a centre-left party does seem to have shaped political elites' preferences on immigration to some extent and thus can shape public policy (Lahav 1997). While the election of a centre-left government was a necessary condition for policy change, it cannot be said to be a sufficient explanation.

The modernisation of the Labour Party proved to have a powerful effect on the preferences of leading political elites and is therefore a significant explanatory factor for why policy developed in the way it did. Labour's rebranding and ideological reorientation through the Party's infamous 1990s policy review were, in part, for electability considerations. Nonetheless, Labour's ideological modernisation was reflected in most public policies, including immigration policy. A major component of the Party's modernisation was their acceptance and endorsement of globalisation. The rebranded Labour Party assumed that globalisation was both inevitable and an intrinsically positive thing. And with international migration being 'the human face of globalisation' (OECD 2009), immigration was, by extension, likewise assumed to be inevitable and primarily positive. These ideas were embraced by the leading factions of the Labour Party and took on an almost unquestioned logic by the early 2000s. In conjunction, Labour's third-way framework meant an overhaul of economic policy, moving away from Keynesian ideas and instead adopting neo-liberal economic principles. A key tenet of this new business-friendly approach was labour market flexibility, and immigration was seen as a tool to aid such flexibility. The Party also espoused an inclusive and cosmopolitan notion of British national identity, and multiculturalism was the preferred framework for integration policy. These principles were compatible with an expanding immigrant population.

However, there was not a homogenous consensus in the Party. Conversely, there was intra-party conflict, with factions condemning some of the policies because of the disproportionate impact on particular communities, an intra-party conflict that continues today. This conflict reflects the contradictory ideological pulls immigration presents to social democratic parties, reflected in the cosmopolitan liberal elite leading the Party, and, according to some members of the PLP, their core constituents' concerns for labour market protectionism. This was especially acute in Labour's final term of office. It was the leading factions of the Party advocating these changes. Nonetheless, if the Party had retained their so-called "Old Labour" ideas of protectionism, it is likely that the status quo of restriction would have persisted. Labour's modernisation agenda, chiefly the Party's acceptance and endorsement of globalisation, was one of the principal causes of the policy change because it changed the preferences of leading Labour ministers.

A weak opposition from the Conservative Party also left the governing Labour Party with a sufficient degree of autonomy. The Conservative Party was fairly passive in their opposition to the policy reforms in the late 1990s and early 2000s. This can partly be attributed to the Party's generally 'futile period in opposition' between 1997 and 2001 (Collings and Seldon 2001, 624). In terms of immigration politics, the focus in the political debate (and the media) was on the 'asylum crisis', and consequently the reforms under study received little political and media attention. To the extent that the policy reforms were debated, the Conservative Party were contrarily fairly supportive. The opposition took a tougher stance on immigration in the 2005 General Election, and the Party has continued to hold a restrictive position on immigration ever since. In turn, Labour began to 'row back', at least rhetorically, on their expansionary approach after the 2005 Election. Yet when the policy reforms were underway between 1997 and 2005 Conservative opposition was weak. If the Conservatives had been stronger in their opposition perhaps the changes would not have occurred so rapidly and with such ease.

A lack of an electorally successful populist far right party also partly explains why the policy reforms transpired without political contention. Far-right parties were becoming relatively electorally successful across many Western European states at this time, and their gaining popularity had an impact on the immigration debate, and in turn policy, in these countries (see Koopmans et al. 2012). This void in the political space meant that the immigration debate was perhaps less fuelled by populism and anti-immigrant sentiment than it otherwise might have been. However, weak opposition and a lack of populist far-right party only partly explain *how* the policy change transpired; it does not provide an explanation for *why* the Labour government passed the policies in the first instance.

Parties do matter then (Schmidt 1996). The modernisation of the Labour Party proved to have a powerful effect on the preferences of certain Labour ministers. Immigration was no longer seen as a threat to national workers' wages. Conversely, immigration was seen as part and parcel of this new globalised economy and a tool to aid labour market flexibility. These ideas permeated through the leading factions of the Party and fundamentally changed their immigration policy preferences. However, these ideas permeated inadvertently. That is to say that Labour did not enter office with a plan to overhaul immigration. Rather the logical extension of an ideology underpinned by economic neoliberalism and the raison d'être of globalisation was an open immigration policy, but this was never formally decided or debated. In this sense, policy change was an unintended outcome of intended action.

The 'parties matter' school of thought assumes that parties are primarily office seeking and therefore adapt their policies in line with the electorate's preferences in a bid to maximise their vote share. Given that the British public have never been in favour of expanding immigration, this case study is particularly illuminating as the policy reforms cannot be attributed to a vote-maximising strategy. Party ideology on the other hand has been neglected as an explanatory factor for why parties formulate their immigration policies in the way they do, especially for social democratic centre-left parties, which have been somewhat overlooked in the literature (see Hinnfors et al. 2012; Odmalm 2014 for exceptions). This research shows that party ideology shaped Labour's economic immigration policy; indeed this was a major cause of the policy shift. It is not simply that 'parties matter' (Schmidt 1996) but rather that party ideology matters.

Labour's rowing back on such reforms after the 2005 General Election can however be attributed to party strategy. This implies that future analyses on the relationship between parties and their influence on immigration policy should take a two-pronged approach to the question, where party strategy is examined but that room is also 'made in explanatory narratives for the role of party ideology' (Hinnfors et al. 2012, 586). Future analyses should not just question whether parties matter but rather show how they matter. This being said, party ideology may be more significant in majoritarian systems where typically one governing party rules, as opposed to pluralistic systems where coalition governments will likely dilute the role of ideology in policy outcomes. Nonetheless, 'in the UK's majoritarian democracy policies move sometimes in an expansionary direction, sometimes in a restrictive direction; and which party or parties are in government matters for explaining this' (Hampshire and Bale 2015, 25).

7.2.3 Administrative Context: Institutions, Policy Frames and Knowledge

Chapter 6 examined the institutions of immigration policymaking. Adopting a disaggregated view of the state the chapter examined the administrative context and the policymaking process in detail, assessing the degree to which policy framings, institutional cultures and institutional changes to policymaking were contributory factors for the regime change. Taking a historical institutionalist lens, the chapter considered policy framings by examining the administrative context of immigration policymaking, such as which departments were involved in the policymaking process and whether this was reorganised under the Labour administration.

There was a stable policy frame underpinning the immigration policymaking in Britain throughout the post-war period. This policy frame was, to some extent, renegotiated in the realm of economic immigration under the Labour administrations. Throughout the post-war period there was a bipartisan consensus that good race relations required limited immigration. Consequently, the policy frame, which had been institutionalised in the Home Office, was that immigration had to be controlled, limited and kept to a minimum. This policy frame, coupled with the other populist policy areas in the Home Office remit, created a cautious departmental culture that was reflected in immigration policy throughout the 1970s, 1980s and 1990s. Immigration was framed as an enforcement issue in need of controlling, much like crime. Asylum policy continued to be couched in a control frame under the Labour administrations. In contrast, economic immigration took on a very different framing.

Under the Labour administrations, immigration, as one homogenous policy, was disaggregated into asylum on the one hand and labour or economic immigration on the other. The Labour Party did not necessarily enter office with a plan to separate these areas. It was rather a cumulative consequence of departmental agendas calling for a need for immigrant labour in their policy remits, which effectively separated these policy streams, along with Blunkett's strategy to appease public concerns over asylum. This separation, consciously or not, proved to be advantageous for the Party, as the government could emphasise control on the one hand in asylum policy 'to allow continuity on the other (labour migration)' (Mulvey 2011, 1478).

The final product of managed migration was a consequence of a series of reforms that stemmed from departments other than the Home Office. Each policy scheme had its own drivers that were deemed politically unproblematic at the time. Combined, these policy schemes produced a liberalising effect on immigration policy. The concept of managed migration provided a cohesive narrative for the combined policy reforms that were already underway and by 2004 liberalised economic immigration, as an economic good, had taken on its own unquestioned logic. The most influential department, and arguably the innovators for this new approach to immigration policy, was the Treasury. For the Treasury immigration was a marginal component of the economic growth strategy.

The challenge for the historical and constructivist institutionalist is not to show that ideas matter, but to explain why an idea is seized at a particular moment. The answer lies in the political and economic context and Labour's institutional reforms to policymaking practices. The strong economic conditions and in turn skill shortages explain why the Treasury were predisposed to encourage expansive immigration policies. Labour's neo-liberal economic framework, coupled with the strong economic conditions, explains the motivations behind the Treasury programmes. The Treasury, with Brown as chancellor, were also especially powerful under the Labour administrations, giving the department more authority and scope to legitimately intervene in public policy than had previously been the case.

The change in policy frame was not so much about new ideas challenging the existing paradigm. Rather, the new frame materialised as a result of spill-overs from one policy subsystem to the immigration policy subsystem, the notion being that immigration was part and parcel of economic policy, and therefore immigration policymaking should be guided by economic efficiency, not necessarily by the electorate's preferences. One could question whether this is a case of a 'third-order change' then (Hall 1993). But given that the reforms were driven by a collective logic, and that such reforms shifted the framing of policy to economic utilitarian underpinnings, I would concede that this case still represents an atypical third-order change, albeit an unplanned one. However, that such change was driven by spill-overs or a collection of 'normal policy changes' demonstrates that an overarching shift can be caused by an accumulation of minor changes or 'layering' (Streeck and Thelen 2005, 22), challenging the consistency of the existing policy regime (Howlett et al. 2009, 202).

The historical institutionalist approach does have some explanatory potential in this case of policy change then. There was a policy frame—to minimise and control immigration—that had structured immigration policymaking in the post-war period. The introduction of new ideas by powerful actors—such as David Blunkett, the Treasury and key actors from IPPR—challenged this framing and transmitted an economic framing to immigration policy for the first time. The objectives of immigration policy changed, so that policy was not based on the notion of controlled immigration being good for race relations. Rather, immigration policy was to be based on economic utilitarian arguments (Balch 2010).

However, whilst the historical institutionalist approach has explanatory potential for this case, the ability to transform the policy frame was ultimately attributable to the wider political context. The opportunity for an idea to take hold was essentially due to a change in the political stream (Kingdon 1995), in other words a change in administration. The change in administration brought in a new set of governmental actors and meant that the ground for new ideas was fertile. The opening of a window of opportunity as a result of a change in the political stream accurately describes how the policy transformation came about. While the historical institutionalist approach has relevance to the case, the change in policy frame was fundamentally rooted in a change in administration, and in turn, the modernisation of the Labour Party. That is to say, the policy frame did shift but that the ability to alter the policy frame and objectives of immigration policy was by virtue of a change to the governing party. This is a challenge to new institutionalist approaches generally, as it demonstrates that far from being 'frozen residual structures' (Thelen 2004, 8), screened from political pressure (Boswell 2007, 83), institutions are in fact susceptible to party political influence and are the 'objects of on-going political contestation' and changes in the political administration 'on which institutions rest are what drives changes in the form institutions take and the functions they perform in politics and society' (Thelen 2004, 31). Thus institutions (in this context at least), whilst important structures that condition policy framings and indeed policy implementation, are essentially political constructions that, although to some extent autonomous and resistant to change, are adaptive and susceptible to political influence from the governing party.

7.3 Pulling the Threads Together: Complex Causality

While the expansionary developments were not caused by one single factor, a set of favourable conditions allowed for a shift in immigration policy. First, a change in administration brought a window of opportunity for policy change (Kingdon 1995). Second, the UK's strong core executive in comparison to other Western states means that the executive can act relatively unconstrained. Consequently if political elites have the political will to change policy then they can do so with relative ease. Third, the 1997 General Election saw a heavy Conservative defeat, leaving the opposition Party deeply divided, uncoordinated and lacking clarity on policy. In turn, the Labour Party won a landslide victory in 1997 making the governing party especially powerful and able to govern with a sufficient degree of autonomy. Fourth, a booming economy meant that the governing party could flex their political muscle freely. These were the favourable conditions—or mechanisms and capacities (Bennett and Elman 2006, 457)—that made the radical shift in policy objectives possible.

Process-tracing Labour's economic immigration policy shows that the regime transformation was an incremental process, albeit over a short period, where initially there was no conscious effort to transform policy. The story of an overarching shift in the policy framework is best seen as an accumulation of a handful of moving independent components, oscillating in the background until, with pressure from the Treasury and a new Home Secretary, the Home Office re-evaluated the system and devised a policy and term that made sense of these reforms—managed migration. What began as a handful of non-contested schemes turned into an immigration system underpinned by the belief that immigration was inevitable, intrinsically positive and imperative for Britain's economic growth. A set of favourable conditions was needed; the administrative context (chiefly the Treasury's early interest) explains the origins, but underlying all of these moving elements was Labour's modernisation project, believing in globalised free markets and friendly capitalism.

The process of a regime transformation was not a simple linear one. There were a number of simultaneous moving pieces in this narrative, all of which contributed to the shift. Thus policy change was ultimately a consequence of favourable timing and complex causality. It is not just the banal truth that context matters in the sense that detail matters. Rather there were different layers of this process, different domains of politics—

non-governmental interests, party politics and the administrative context of policymaking—that move according to their own logic. Public policymaking brings these layers to the fore, and policy change can be a reflection of these different interacting layers: contingency, sequencing and timing matter (Pierson 2004). This is not to suggest that each social setting is unique and infinitely complex, but that we 'should recognize that an event or process is environed by its temporal location, its place within a sequence of occurrence, and by its interactions with various processes unfolding at different speeds' (Pierson 2004, 172). If, for example, the economic climate had been poor, the policy reforms would not have materialised. Likewise if Labour had not ideologically re-orientated in the first instance, there would have been no incentive to change the status quo of restriction. And without administrative reforms opening up the 'turf' for other departments and actors, and an especially powerful Treasury under Brown, Labour's ideology would not have been exported to the framing of immigration policy.

New Labour was a new breed of politics; it was a very different beast from its predecessors and a powerful one at that, both directly in terms of their majority, the personalities of the core leadership, but also rhetorically and discursively. Successful or not, the third way was a genuine attempt to redefine British politics, and Blair has been described as the 'master of new politics' (Gamble 2010, 645). This was said to be 'the party of change, of progress of radicalism' (Robinson 2012, 123) and Labour's dominance and ideological persuasion are at the heart of this case.

This case in many ways resembles a critical juncture then, as this set of favourable conditions provided a window of opportunity in which structural influences on political action were relaxed. These conditions—strong economy, strong majority government and a weak opposition—coincided at the same time and consequently prompted a moment for political actors to overcome the usual bias towards inertia (in this case a restrictive policy frame). From this perspective, one could say that three variables converged to produce the liberalisation of immigration: a strong economy with low unemployment and skills shortages; a Labour government with a colossal majority and an ideological commitment to globalisation and free markets; and a series of institutional reforms that broke the monopolisation of the Home Office by introducing new actors into the immigration policymaking process. Independently, none of these factors would likely have been sufficient to cause such a radical change. A strong set of structural conditions (strong economy, strong governing party) created a window of opportunity and allowed for the contingent combination of these factors (Labour's third-way ideology, and change to the machinery of government, which broke the Home Office monopoly) to displace the typically restrictive policy framing (Consterdine and Hampshire 2014).

7.4 Explanatory Insights for Policy Change

If we want to draw lessons on why policy changes, four major findings stand out. First, there was no single cause of change. Public policymaking is often a chaotic and convoluted process, and conditioned by a multitude of factors, including socioeconomic conditions, the governing party, departmental agendas, public opinion, knowledge and evidence and organised interests to name a few. This case highlights the limitations of the literature on immigration policymaking, which privilege one factor over another, such as political economists' preoccupation with interest groups, party scholars' concentration on political parties and the institutionalists' focus on the administrative context, all at the expense of recognising the contingency between these factors. A single causation of policy change does not reflect the reality in this case. As Pierson (2004, 178) argues, 'all angles of vision create distortions'; thus scholars should resist the temptation of adopting a single lens to explain public policymaking. Those searching for parsimonious explanations will fail to recognise the non-trivial complexity of policymaking and the interplay between conditions such as party politics and the institutional context, where the explanation of policy change often lies.

Second, in the context of immigration policymaking at least, executive power is alive and well. Policy change was driven by political elites and was executive led. The UK is the Weberian ideal type of a strong executive and, as Hansen has noted, this means that the UK can maintain 'for relatively long periods policies that fly in the face of public opinion' (Hansen 2011, 23). Hansen (2000) has argued elsewhere that the turn to restrictive immigration policies in 1960s Britain was attributable to a strong core executive. Forty years later this pattern of concentrated executive power remains in place. Under favourable conditions immigration policy is determined 'top down' in a relatively autonomous way by the executive in Britain and, in turn, by the leading political elites.

A further lesson is the potential influence that party ideology can have on policy preferences. The scholarly literature tells us that party affiliations make a limited difference to elites' immigration policy preferences. This book challenges this assumption. Explanations of immigration policy change have tended to focus on ideas, interests and exogenous factors at the expense of the party political context. Parties are too often assumed to be passive, merely channels for other actors preferences and ideas, changing policies to appease the electorate or simply reacting to exogenous events. Conversely, those that have studied the relationship between parties and immigration policy have focussed exclusively on party positioning and party strategy (see Hinnfors et al. 2012 and Odmalm 2014 for exceptions). Yet in this case the logic behind managed migration was a reflection of Labour's distinct third-way vision of the global economy. An ideological reorientation in a party can fundamentally alter the preferences of political elites, and if this party enters government this reorientation will likely be reflected in policy and, if deemed necessary, will result in policy change. This book also demonstrates that the governing party ideology can be reflected in administrative reforms and thus influence the institutional context of policymaking. Intra-party change, such as an ideological shift in a party, can potentially have a major impact on the direction of policy then. Not only do parties matter, then, but party ideology matters, and future analyses should examine both party strategy *and* party ideology.

A final lesson is the importance of policy spill-overs and therefore the importance of studying policy developments in other policy subsystems. The impetus for a managed migration policy was a result of several policy reforms stemming from other policy agendas. In the making these reforms were not conceived as one homogenous policy programme. A shift in the policy regime is not necessarily a conscious political decision and can be a repercussion of other policy choices and outputs. This further highlights the need to take a disaggregated view of the state, lest the decision-making processes will be lost. Scholars should resist studying immigration policy in an enclave, detached from other policy agendas. Immigration policy can be informed by a number of objectives, including foreign policy, higher education policy and economic policy. Indeed, in this case many of the reforms were a by-product of other policy agendas. Both those who study immigration policy and those who seek to explain policy change more generally have neglected the impact of policy spill-overs. Future analyses of policy change should give some consideration to policy developments in other sub-systems that may be influencing the direction of the policy area under question.

Since labour left office in 2010 a watershed moment has come to pass, one that arguably reflected a rejection of Labour's managed migration regime — in June 2016 the UK public voted to leave the EU — a seismic shift that will shape Britain's foreign relations and domestic politics for years, if not decades, to come. Immigration is now at the top of the political agenda; the issue is undoubtedly highly salient, polarising and politicised. To conclude the book we reflect on how we got to this stage and what Labour's legacies have been on the politics of immigration and beyond.

REFERENCES

Balch, A. (2010). *Managing Labour Migration in Europe: Ideas, Knowledge and Policy Change*. Manchester: Manchester University Press.

Bennett, A., & Elman, C. (2006). Complex Causal Relations and Case Study Methods: The Example of Path Dependence. *Political Analysis, 14*(3), 250–267.

Blunkett, D. (2003). *Towards a Civil Society*. London: IPPR.

Boswell, C. (2007). Theorizing Migration Policy: Is There a Third Way? *International Migration Review, 41*(1), 75–100.

Collings, D., & Seldon, A. (2001). Conservatives in Opposition. *Parliamentary Affairs, 54*(4), 624–637.

Consterdine, E., & Hampshire, J. (2014). Immigration Policy Under New Labour: Exploring a Critical Juncture. *British Politics, 9*(3), 275–296.

Gamble, A. (2010). New Labour and Political Change. *Parliamentary Affairs, 63*(4), 639–652.

Hall, P. (1993). Policy Paradigms, Social Learning, and the State: The Case of Economic Policymaking in Britain. *Comparative Politics, 25*(3), 275–296.

Hampshire, J., & Bale, T. (2015). New Administration, New Immigration Regime: Do Parties Matter After All? A UK Case Study. *West European Politics, 38*(1), 145–166.

Hansen, R. (2000). *Citizenship and Immigration in Post-war Britain*. Oxford: Oxford University Press.

Hansen, R. (2011, May 18–21). *Paradigm and Policy Shifts: British Immigration Policy, 1997–2011*. Paper Presented at the Conference on 'Controlling Immigration: A Global Perspective', The Federal Reserve, Dallas.

Hinnfors, J., Spehar, A., & Bucken-Knapp, G. (2012). The Missing Factor: Why Social Democracy Can Lead to Restrictive Immigration Policy. *Journal of European Public Policy, 19*(4), 1–19.

Howlett, M., Ramesh, M., & Perl, A. (2009). *Studying Public Policy: Policy Cycles and Policy Subsystems*. Oxford: Oxford University Press.

Kingdon, J. (1995). *Agendas, Alternatives, and Public Policies*. New York: HarperCollins.

Koopmans, R., Michalowski, I., & Waibel, S. (2012). Citizenship Rights for Immigrants: National Political Processes and Cross-National Convergence in Western Europe, 1980–2008. *American Journal of Sociology, 117*(4), 1202–1245.

Lahav, G. (1997). Ideological and Party Constraints on Immigration Attitudes in Europe. *Journal of Common Market Studies, 35*(3), 377–406.

Mulvey, G. (2011). Immigration Under New Labour: Policy and Effects. *Journal of Ethnic and Migration Studies, 37*(9), 1477–1493.

Observer. (2003, September 7). Matthew Taylor Profiled. *Observer*. Available from: http://www.theguardian.com/politics/2003/sep/07/society.thinktanks. Accessed 7 Mar 2012.

Odmalm, P. (2011). Political Parties and "the Immigration Issue": Issue Ownership in Swedish Parliamentary Elections 1991–2010. *West European Politics, 34*(5), 1070–1091.

Odmalm, P. (2014). *The Party Politics of Immigration and the EU*. Hampshire: Palgrave.

OECD. (2009). *International Migration: The Human Face of Globalisation*. Paris: OECD.

ONS (Office National Statistics). (2012). *2011 Census, Key Statistics for Local Authorities in England and Wales*. Available from: http://www.ons.gov.uk/ons/rel/census/2011-census/key-statistics-for-local-authorities-in-england-and-wales/index.html. Accessed 11 Nov 2013.

Pierson, P. (2004). *Politics in Time: History, Institutions, and Social Analysis*. Princeton: Princeton University Press.

Robinson, E. (2012). *History, Heritage and Tradition in Contemporary British Politics: Past Politics and Present Histories*. Manchester: Manchester University Press.

Sassen, S. (1999). Beyond Sovereignty: De-facto Transnationalism in Immigration Policy. *European Journal of Migration and Law, 1*, 177–198.

Schmidt, M. (1996). When Parties Matter: A Review of the Possibilities and Limits of Partisan Influence on Public Policy. *European Journal of Political Research, 30*(2), 155–183.

Spencer, S. (1994). *Strangers and Citizens: A Positive Approach to Migrants and Refugees*. London: IPPR/Rivers Oram Press.

Spencer, S. (2007). Immigration. In A. Seldon (Ed.), *Blair's Britain 1997–2007* (pp. 341–360). Cambridge: Cambridge University Press.

Statham, P., & Geddes, A. (2006). Elites and the "Organised Public": Who Drives British Immigration Politics and in Which Direction? *West European Politics, 29*(2), 248–269.

Streeck, W., & Thelen, K. (2005). Introduction: Institutional Change in Advanced Political Economies. In W. Streeck & K. Thelen (Eds.), *Beyond Continuity* (pp. 1–40). Oxford: Oxford University Press.

Thatcher, M. (1992, May 15). *Speech in the Hague ("Europe Political Architecture")*. Available from: http://www.margaretthatcher.org/document/108296. Accessed 15 May 2012.

Thelen, K. (2004). *How Institutions Evolve*. Cambridge: Cambridge University Press.

CHAPTER 8

Beyond New Labour

Seven years since Labour left office and now immigration is front and centre of the British political landscape. For over 50 years' politicians had largely managed to duck the issue and now no party can afford to ignore it.

As we have seen, Labour rowed back on their retrospective regime in their final term under Prime Minister Brown, best encapsulated with his call for 'British jobs for British workers' in 2007. Nonetheless, Labour's expansive immigration regime was definitively scrapped when they lost the 2010 General Election, and a new Conservative-led Coalition government entered office with a pledge to reduce net migration from the hundreds of thousands to the tens of thousands by the end of the Parliament (Conservative Party 2010).

There are a number of reasons why Labour lost that election (*see* Geddes and Tonge 2015), but paramount was the Conservative's successfully mobilising the narrative that Labour had caused the 2008 global financial crash, 'and to heap all the blame for the [...] ensuing recession on to Labour's mismanagement of the public finances and of the wider economy... The crisis was framed as the latest example of Labour's economic incompetence' (Gamble 2014, 6). A narrative that was so successful, it likewise tainted Labour's branding of economic competence in the 2015 General Election (Cowley and Kavanagh 2016). Nevertheless, the Party's apparent mishandling of immigration was also a contributory factor to their 2010 defeat (Evans and Chzhen 2013; Bale 2014) and one that has continued to dog the Party ever since.

© The Author(s) 2018
E. Consterdine, *Labour's Immigration Policy*,
https://doi.org/10.1007/978-3-319-64692-3_8

Few would deny that immigration has become a more salient issue in Britain. The percentage of voters who wanted to see 'immigration reduced' rose from 39 per cent in 1995, to 55 per cent in 2008, reaching a peak of 56 per cent in 2013 (Dennison and Goodwin 2015, 175; British Social Attitudes survey). Whilst the British public has long been in favour of reducing immigration, the high level of public concern has been more recent, gravitating from a marginal concern of a small minority to what voters considered as one of the most important issues facing Britain (Ipsos-Mori 2015; Duffy 2014). The heightened public concern around immigration began in 2000, as we have seen at a time where the New Labour government was pursuing the most expansive immigration regime to date (Ipsos-Mori 2015).

Increases in actual numbers do not always drive public anxieties over immigration (Ivarsflaten 2005) and indeed those who are most concerned tend to reside in areas least affected by immigration (Ford 2011). But whilst 'it is impossible to entirely separate the importance of media and political narrative from actual changes in numbers…the relationship is clear enough to conclude that the number of immigrants is important to public attitudes, and particularly to how salient they feel the issue is' (Duffy 2014, 259–60). And although media and political party portrayals of immigration undoubtedly shape public opinion, 'rises in immigration are associated with increased demands for restriction' (Ford et al. 2015, 1397) Further research also suggests that the speed of ethnic change because of immigration, as opposed to the stock of immigration per se, plays a major role in driving anti-migrant sentiments (Kaufman and Harris 2014; Portes 2016, 105; Coyle 2016, 25; Clarke and Whittaker 2016). As we have seen, such rapid changes in the demographic population were largely due to Labour's policies.

The intensified political debate over immigration is likewise hard to deny. By way of illustration, whilst Blair made only two speeches on immigration over the course of his ten-year premiership, in contrast David Cameron delivered four—almost a speech for each year as Prime Minister. Immigration barely warranted a mention in party manifestos until 2010, yet now every office-seeking party must have a clear policy on immigration. For example, whilst the Conservative's 1997 Manifesto contained two sentences on immigration and asylum, the 2015 manifesto contained a whole chapter.

And of course, a seismic shift in British politics, and a decision that will likely define generations to come, has transpired since Labour left office—

in June 2016 Britain voted to leave the European Union, a vote that was fuelled by anti-immigration and anti-establishment feelings (Hobolt 2016, 1260). For arguably if Britain's vote to 'leave the European Union was a vote against anything, it was a vote against free movement of workers within the EU—a vote to "take (back) control" over immigration policy' (Portes 2016, 105). At the very least, this is how Prime Minister Theresa May and the majority of parliamentarians (Ipsos-Mori 2017) interpreted the result, explicitly stating she would prioritise controls over immigration over access to the single market (May quoted in BBC 2017).

The question is how and why has immigration gravitated from low to high politics? And what were the consequences of the New Labour period? This chapter reflects on what Labour's legacy has been on the policy and politics of immigration in Britain and the repercussions of the Labour's immigration regime on British politics. The chapter reflects on the state of play and radical developments in immigration politics in Britain from when Labour left office in 2010 until the June 2017 General Election.

The chapter is divided into three sections. Previous policy decisions can lay a particular path for future policymaking and policy outputs, which are referred to as feedback effects. The chapter begins by briefly assessing the feedback effects of New Labour's expansive regime on the institutions of immigration policymaking and organised interest group formation. At the same time, 'polices can create new politics' (Schattschneider 1935) by influencing political actors, organising political understandings and structuring political relations (Moynihan and Soss 2014, 3). The second section of the chapter charts the quite radical changes in party political positions and dynamics since 2010 and links such developments to the 2004 A8 decision. The final section concludes the book by returning to Labour's legacies on the politics of immigration and beyond, where above all else Labour's regime has brought immigration to the forefront of the political landscape.

8.1 Feedback Effects: State Enhancing and Interest Group Formation

Britain's majoritarian system and strong core executive mean that governing parties can shift policy radically if they deem it necessary, and the Conservative-led governments clearly did. Entering office with a pledge to reduce net migration (Conservative Party 2010)—a pledge the

Conservative-led administrations have consistently failed to meet but continue to retain (Conservative Party 2017)—restrictive measures over the period 2010–2017 amongst others included closing Tier 1 general and Tier 1 post-study work; placing a cap on the number of Tier 2 visas issued annually; closing both the Seasonal Agricultural Workers Scheme and the Sector Based Scheme; and a swathe of draconian measures on language requirements, income thresholds, economic resources and increasing settlement requirements across all migration streams (Gower 2015; Home Office 2015)). At first glance it might seem that Labour's expansive regime was a deviation then, confined to its time and quickly dismissed as a 'mistake' in British politics (Miliband 2012a). Yet the changes made have had feedback effects on both the way immigration policy is structured and the wider politics of immigration, which is a fundamental, if overlooked, legacy of the New Labour period.

The first feedback effect stems from the preservation of the points-based system (PBS). As the Minister of State for Immigration between 2010 and 2012 Damian Green acknowledged, the government chose to 'keep the framework of the migration system created by the last Government and focus on getting the detail right' (Green 2012). This generated both a positive self-reinforcing state-enhancing capacity (Skocpol 1992) effect and an unintended negative feedback effect in the way of policy implementation.

The maintenance of the PBS allowed the government more control over immigration admission in a highly flexible way, whilst at the same time outsourcing the costs of implementation and liability on to employers and universities, producing a state-enhancing feedback effect. As a former Home Office SpAd commented 'it was a system which had been designed deliberately so that you could make adjustments to it' (Interview with former SpAd 2012), particularly advantageous in light of Brexit negotiations, and the likely necessity to regulate and therefore possibly absorb EU labour migration into the system.

Whilst the PBS has strengthened state capacity in the area of migration management, the most significant feedback effect of the system has been an unintended effect on policy implementation practices. As one interviewee put it, the PBS represented the 'big bang moment…the biggest change in immigration policy in 40 years' (Interview with UKVI, 2015). The system has evolved from highly discretionary in the 1990s, to limited discretion under the PBS, but as a result of the unintended outcomes from the necessary blanket approach of the PBS, the Home Office's wing for

8.1 FEEDBACK EFFECTS: STATE ENHANCING AND INTEREST GROUP... 223

visas and immigration—UK Visas and Immigration (UKVI)—are now moving back towards introducing formal discretion. Whilst the PBS was in part set up to make the system objective and transparent, due to complaints of inconsistent decision-making from the discretionary work permit system, this meant that, as one UKVI interviewee described, some applicants 'met the criteria but not necessarily the spirit or intent of the policy'. In short, whilst applicants could technically fulfil the criteria, they were not necessarily the applicants the government is trying to attract. As a result, UKVI have moved towards reintroducing formal discretion, including video interviewing (via Skype at Global Hubs), credibility testing, genuine vacancy testing and a policy of evidential flexibility, moving the system towards a subjective caseworker model.

The second feedback effect of Labour's managed migration programme relates to its impact on interest groups and their role in policymaking. The establishment of the Migration Advisory Committee (MAC) under the Labour administration discussed in Chap. 4 has made the government dependent on employers for labour market evidence in order to design policies to meet market demands, shifting both the incentives and resources of interest groups. Thus interventionist policies have had the adverse effect of making policy, and therefore the government, dependent on interest groups, in turn creating what Caviedes (2009) calls an 'indirect corporatist agreement'.

Whilst the MAC was originally on a register of quangos to be abolished as the Coalition government entered office in 2010, a Cabinet Office review concluded that the body should be maintained on the grounds that it was 'performing a function which requires impartiality' (Cabinet Office 2011). Employers have also called for the MAC's role to be more extensive by 'overseeing the administrative impact of immigration policy' (CIPD 2013). By February 2017, the MAC had published no fewer than 67 reports, of which the government have accepted almost every recommendation.

Whilst publicly the Conservative-led governments have pursed their policies regardless of business opposition, lobbying has certainly tempered government proposals. For example, the exclusion of intra-company transfers from the annual Tier 2 limit was a result of the intensive and coordinated lobbying from employers and employer associations. As a result of business lobbying, the Coalition government made several amendments to their original immigration plans. For example, whilst an annual quota was established for Tier 2, this was set at a relatively high level of 20,700.

Furthermore, decisions on where restrictions should be placed were made on the basis of employers' consultation responses such as prioritising Tier 2 visas but closing Tier 1 general and retaining a flexible approach for scientists and researchers (Green 2012). Following lobbying from a number of banks and law firms in the City, foreign workers earning over £150,000 were exempt from the annual cap. Additionally, the government in the face of business opposition (BBC 2017) quickly abandoned Home Secretary Amber Rudd's suggestion in October 2016 that employers should have a list of foreign workers. Indeed in the face of business opposition to the prospect of losing EU labour, the government has indicated that they may re-establish the SAWS (Leadsom quoted in Independent/Cowburn 2017).

The role of business has outlived the specific conditions of 2001–2005, and despite adverse economic conditions following the 2008 financial crash, and a new government committed to reducing net migration, employers and employer associations continue to be influential in immigration policymaking through the MAC and other channels in ways that were unheard of before the early 2000s. Consequently, future governments face greater scrutiny from this emerging policy community, especially in the context of Brexit and the consequential concerns over labour shortages. We now move to the repercussions of Labour's regime on the wider political context, where the 2004 A8 decision triggered a ratcheting effect across the political spectrum.

8.2 The Politics of Immigration: Unintended Developments of the A8 Decision

To understand how and why the once separate issues of EU membership and immigration came to be entangled in public debate, as evidenced by the referendum campaign and result, one must look to the earlier policy decision made by the Labour government.

As we saw in Chap. 3 and in more detail in Chaps. 5 and 6, in May 2004, the EU welcomed ten new member states—the majority from Central and Eastern Europe—in what was the largest expansion in the history of European integration, otherwise referred to as the A8 decision. This was to be perhaps Tony Blair's greatest unintended legacy as the decision had major consequences for both the patterns and politics of migration (Geddes 2014, 290).

The A8 decision goes some way to explaining the fusion of the once separate issues of EU membership and immigration and consequently why much of the electorate voted to leave the Union. As we saw earlier in the book, the decision was uncontroversial at the time and barely framed or regarded as an immigration issue. Yet what we also know is that the government were ill prepared for the magnitude of immigration flows from the CEE states. Whilst Dustmann's infamous report predicting migration flows in the range of 5,000–13,000 had limited bearing on the government's decision, it is nonetheless clear by the Labour Party's own admission that they had not anticipated just how great this immigration would be. In turn, we saw former Labour leader Ed Miliband, as well as former ministers, literally apologising for this decision (Miliband 2012). One of the main protagonists of the New Labour project—former Home Secretary (1997–2001) and Foreign Secretary (2001–2005) Jack Straw—went so far as to state that Labour made 'a spectacular mistake' in not imposing transitional controls on A8 countries (Straw quoted in The Telegraph 2013).

Aside from the physical change this brought to the population, the importance of this decision lies in the fact that it opened the political space for right wing competitors to fuse and own the issue. The government's A8 decision and the speed of immigration this brought now left a political space for opposition to freedom of movement, and with a Eurosceptic fringe party in waiting, the time was ripe for capitalising on this issue. The implications of the A8 decision allowed what had hitherto been an ineffectual pressure group to gain the political ground and fuse the core issue of its campaign—leaving the EU—with immigration, which in turn led to a ratcheting effect across the political spectrum. Enter Farage's UK Independence Party (UKIP).

8.2.1 UKIP and the Fusion Strategy

UKIP of course did not emerge because of the A8 decision. The Party had started as a pressure group against the European project following the Maastricht Treaty in the early 1990s. Ford and Goodwin's (2014) seminal account of the rise of UKIP demonstrates how a change in political strategy transformed this once minor fringe anti-EU pressure group into a major political force. Ford and Goodwin argue that in UKIP's early years, the party focused their electoral efforts on appealing to middle class Eurosceptic Conservatives who were angry following the passing of Maastricht Treaty in 1991. However, after 2009 the party changed tack and began appealing to Labour's core electoral base—disadvantaged, 'white working class people whose traditional loyalty to the centre left had eroded, and who stayed at home on election days or flirted with the extreme right BNP' (Ford and Goodwin 2014, 108). To attract these once Labour voters, UKIP began fusing their hard Eurosceptic message with,

> Stronger nationalist, anti-elitist and anti-immigration elements in the hope of taking back votes from both Labour and working-class Tories, who once backed the assertive nationalism and traditional values of Margaret Thatcher, but who since 2005 have been less receptive to Cameron's compassionate Conservatism. (Ford and Goodwin 2014, 109)

Former UKIP leader Paul Nuttall highlighted the potential to capitalise on Old Labour voters in 2007 when he wrote that there was one powerful issue of which New Labour's actions had enraged former voters: immigration (Ford and Goodwin 2014, 109). These traditional Labour voters, Nuttall argued, were opposed to mass migration, seeing the influx of foreigners as threatening their wages, jobs, identity and cohesion of their communities. This strategy proved effective indeed; in 2016 UKIP had more working class party members than Labour in terms of percentage (Bogdanor 2016, 8).

Whilst the UK's First Past the Post electoral system prevented UKIP's entry into Parliament—as a major party at least—the success of their fusion strategy was evident. First, the Party had a landslide victory at the 2014 European Parliament elections. Second, whilst the electoral system meant UKIP only gained a solitary seat in the 2015 Election (although this was in itself a victory being the first time UKIP had ever held a seat in the Commons), their vote share increased substantially to 12.6 per cent, overtaking the Liberal Democrats in this respect as the third largest party. Finally, arguably UKIP's biggest success was their role, and in particular, that of Farage, in convincing the British public to leave the EU.

UKIP provided a rhetoric that few had previously adopted, one based on above all else reducing immigration and exercised through dog whistle politics. The void of a radical right challenger that aided Labour's ease in passing their expansive programme in the 2000s was now replete. This created a ratcheting effect across the political spectrum, because now the defusion strategy (see chapt 2. and Bale et al. 2010) was redundant. With a party willing and indeed indoctrinating the need for reduced immigration as part of their ideological core, the once unorganised anti-migrant public, which Gary Freeman spoke of, was now organised and mobilised by UKIP.

8.2.2 Tories: Accommodative Strategy

With UKIP now gaining ground, and an anti-immigration stance being the winning ticket, once socially liberal Cameronism (Evans 2008) became increasingly authoritarian in policy and discourse. Whilst the Conservatives have frequently called for curtailments in immigration policy, never has the Party set itself an actual migration target in government. In the same vein, whilst a large faction of the Conservatives had been vehemently Eurosceptic, and various Tory leaders have espoused these sentiments to varying degrees in rhetoric and policy, the Party has never seriously contemplated leaving the EU as party policy. The reason Cameron adopted both the net migration policy and agreed to an in/out referendum—notwithstanding intra-party pressure—was the threat of UKIP.

Cameron adopted an accommodative strategy then (Bale et al. 2010; Bale 2008), in a bid to outdo UKIP, amp-ing up the rhetoric and policy (Norris 2005). Cameron's rhetoric as opposition leader in the late 2000s was fairly nuanced, claiming in 2006 that it was wrong to blame immigrants for driving down wages and that we must be honest about the costs of globalisation, 'because the alternative is that people project their fears and anxieties on to other ethnic groups or other countries' (Cameron quoted in BBC 2006). Yet by the time the Conservatives took office, Cameron was very much framing increasing immigration as a problem, claiming towards the end of his premiership that 'if you have uncontrolled immigration, you have uncontrolled pressure on public services... Uncontrolled immigration can damage our labour market and push down wages...[And that he] and many others believe it is right for us to reduce the incentives for people who want to come here' (Cameron 2015).

Despite the swathe of restrictive policy reforms to curb immigration, by the 2015 Election campaign the Conservatives were in a tight spot; polling was unequivocally showing the need for the Party to respond to people's concerns over immigration, but they were running out of migration channels to cut. The restrictive measures had failed to win them many points at the booths, with polling indicating increasing discontent with immigration, although the Tories still comfortably owned the issue, being at a minimum 12 points ahead of Labour on immigration throughout

2010–2015 (Dahlgreen 2013). Despite net migration still running high, according to Lord Ashcroft's 2015 polling, 'the Tories were thought more likely than Labour to bring about "a significant fall in immigration to the UK"' (Ashcroft 2015), which is perhaps more a reflection of the publics' distrust with Labour on immigration than confidence in the Conservatives.

Nonetheless, the government had set itself a target, a very explicit and measurable one, and as a result, the Office for National Statistics (ONS) quarterly statistics were now making headlines. Over the five years under the Coalition administration, the net migration figures made 84 headlines across the main national newspapers. We have seen in Chap. 6 what happens when governments play the numbers game, even more so when they cite an actual number to (albeit inadvertently) hold themselves to account. What began as an electoral strategy to win votes then became a measurable instrument of policy failure.

The government had to somewhat concede they failed, but to not lose too much face happily shifted the blame to the fact of the UK's membership of the EU and therefore technically uncontrolled freedom of movement. In turn, Cameron claimed that 'to truly succeed in controlling immigration, we also need the third part of this approach: reducing the incentives for people coming here from within the EU' (Cameron 2015) and that he therefore 'wanted to create the toughest system in the EU for dealing with abuse of free movement' (Cameron 2014).

The government's persistent framing of immigration as a problem in need of controlling had only served to reinforce in the publics' mind that immigration was out of control and ratchet the issue up where 'increased salience leads to increased attention that, in turn, increases the number of government statements emphasising the negativity of migratory movements' (Bale et al. 2008). With the link now firmly established between EU membership and increasing immigration, UKIP had extra political capital to criticise and lobby against Britain's EU membership. Of course what the Conservatives failed to anticipate, so sure was Cameron of his decision to pledge a referendum to win the 2015 Election, was that a sizeable chunk of the public would express their anger over what they considered uncontrolled immigration by voting to leave the Union.

8.2.3 Labour and Immigration: The Elephant in the Room

What of the Party that brought these fundamental changes to immigration policy? For the shadow Labour Party, how to handle the issue of immigration has proven a backbreaking task for both the former, Ed Miliband, and, at the time of writing, current (Jeremy Corbyn) leaders. The ideological tensions immigration presents to centre left parties have come to the fore and the Party is split, an exemplification of the wider ideological problem that the issue forces the Party to confront and one that remains very much unresolved. What this has meant for the politics of immigration in Britain is a further ratcheting effect on the saliency and, in turn, a new bipartisan consensus that substantive reductions in immigration are imperative.

Ultimately, whilst Ed Miliband's 'background and instincts [we]re those of a metropolitan liberal' (Hardman 2014) and the Party's values are founded on notions of equality, tolerance and openness, all of which lend to a progressive immigration policy, at the same time part of the Party's traditional working class electoral base were clearly demanding less immigration. Miliband knew this dilemma as leader, declaring the two dangers of the immigration debate to be either to wish away public concerns or to suggest that Labour could close Britain off from the world, a difficult balance that he acknowledged to be 'an incredibly hard thing to achieve' (Miliband 2012b). Above all else, the Party knew that they had to disassociate themselves with their previous record in office if they were to win on an issue that the Tories historically 'own' (Green and Hobolt 2008), let alone win back voters that had, or would, defect to UKIP (Cruddas 2016). For Labour's first term in opposition, immigration seemed to be the elephant in the room, an issue for which the Party seemingly could not square the circles.

What resulted was an immigration policy under Miliband that to all intents and purposes mirrored the Conservative's bar to a EU referendum pledge. The substance of the policy came in the run-up to the 2014 European elections, with Labour much like their Tory counterparts looking over their shoulders to an increasingly popular UKIP. Shadow Home Secretary Yvette Cooper gave a series of speeches in the lead up to the elections, with a steady drip of restrictionism in tone and policy. She tried, although seemingly failed, to carve a narrative that Labour's immigration policy was both different from the Tories (all immigration is bad) and the free market liberal approach adopted by their New Labour predecessors (most immigration is good). Although Miliband had initially been reluctant (Miliband 2012b), Cooper committed Labour to the Tories Tier 2 cap and set a pledge that under Labour 'all public—sector workers who work directly with the public will have to be able to speak English to a decent standard' (Cooper 2014).

Yet like their governing opposition, by echoing the tough approach of restriction, the Party adopted a counterproductive position, whereby yet again they served to ratchet the issue up, 'boosting Farage's party by helping to ensure that immigration remained at the top of the public's list of "most important issues facing Britain"' (Bale 2015, 214).

Notwithstanding the wider contention of Corbyn's leadership, Labour's immigration policy is still ambiguous and the wider ideological tensions, which form a core theme of this book, remain unresolved. On the one hand, Corbyn's rhetoric is progressive; he claims Labour will not 'sow division or fan the flames of fear' (Corbyn 2016) by promising to cut immigration. His Party will 'instead tackle the real issues of immigration—and make the changes that are needed', such as 'taking action against undercutting of pay and conditions, closing down cheap labour loopholes, banning exclusive advertising of jobs abroad and strengthening workplace protections' (Corbyn 2017). At the same time, those factions of the Labour Party calling for a different, harder, stance towards immigration, now vindicated in their concerns as expressed by losing voters to UKIP, and more importantly a slim majority of the public opting to leave the union, persist with more vehemence than under Miliband. Prominent backbenchers including Rachel Reeves, Chuka Ummuna and Andy Burnham have persistently called on Corbyn for a new debate on freedom of movement within the Labour Party and have stated that the government should seek to stem the flows of EU migrants. In turn, Corbyn instructed his MPs on a three-line whip in January 2017 to vote to trigger Article 50, stating that Labour is not 'wedded to the principle of free movement' (Corbyn 2017).

8.3 New Labour's Legacies

8.3.1 Immigration and Labour's Identity Crisis

At the heart of Labour's current infighting is the theme we came back to time and again in this book—the contradictory ideological pulls that immigration presents to centre-left parties between on the one hand cosmopolitan internationalism and on the other welfare state/labour market protectionism (Odmalm 2014). These tensions were always extant, but before 1997, this did not matter; where once immigration barely warranted a mention, now every party must have a clear policy on the issue. In this respect, a key repercussion of New Labour's reforms is that they have brought immigration to the forefront of British politics and debates on immigration will remain a key fixture of the political landscape.

Labour needs to win back their lost traditional working class voters whilst retaining their liberal metropolitan so-called Blairite base to have any chance of winning office (Fabian Society 2017). The twin pulls of populism and pluralism have restricted Labour parliamentarians, and the Party seem baffled about how to reach the white working class core constituents 'without alienating other crucial segments of their electoral coalitions' (Gest 2016, 19). In short, 'with questions of immigration and identity at its core, the referendum on Britain's membership of the EU inevitably cut across Labour's electoral coalition' (Curtice 2016). Labour needs to appeal to both their (in crass and simplified terms) lost 'leavers' (beer drinkers) and their liberal 'remainers' (wine drinkers). However, appealing simultaneously to the concerns of London voters and the very different interests of voters in the deindustrialised North is a massive challenge and possibly an insurmountable one. It also raises questions surrounding electoral choice in Britain—for what is there to separate Labour from the Conservatives or UKIP if they espouse the same restrictive rhetoric? Moreover, how can Labour outflank the Conservatives on an issue they have long owned, let alone UKIP? This is the progressive dilemma of Labour, one brought to the fore by New Labour's immigration regime, and if left unresolved may signal their demise as a key political force.

Miliband and Corbyn's diagnosis of the publics' anxiety over immigration being ultimately about grievances with the system that does little to serve them may be accurate. But measures to root out the causes of irregular migration, greater employer liability to prevent undercutting and investment in infrastructure, housing and public services to appease these grievances are long-term solutions and ones that the public will not necessarily buy in a country where parliaments are short and fixed. Labour only seem able to offer economic solutions to what it seems is for many a question of culture and identity (Ivarsflaten 2005; Kaufman 2016) in tangent with the broader shift away from the traditional left/right axis and towards post material values (Inglehart 2008). The issue is difficult for Labour precisely because it touches on the nebulous and toxic questions of culture and identity for which they have few answers. Whether economic grievances—egocentric or socio-tropic—do in fact drive public anxieties or whether discontent over immigration is a proxy for other grievances is a question scholars and politicians continue to ponder. The answer may be an uncomfortable one for the Party but is nonetheless key to the kind of policies Labour should offer.

In many ways where the Party is now is a familiar bypass for Labour. As we saw in Chap. 5, in the wilderness years the centre left had to reconfigure, accommodate and realign their ideology with the new interdependent world they faced, or at least claimed they faced. Hence the old protectionist, and to some extent socialist (Leach 2015, 1124), traditions of the Party, which were seemingly not winning voters, were abandoned.

However, whilst Labour's third-way globalist ideology chimed with the times, it was at heart always problematic. The two wings at the core of the third way of liberalism and communitarianism expressed by the central principle of 'no rights without responsibility' played well for the Party. However, in policy practice it expressed the contradictory tenets of this philosophy by trying to preach and privilege both the meritocratic right of individual liberty, whilst at the same time anchoring communities, society and cohesion as the cornerstone of the third way. Blair put forward this somewhat paradoxical vision perfectly at the Labour Party conference in 1994 when he said, 'We are the party of the individual because we are the party of the community' (Blair, Labour Conference Speech 1994). The politics of paradox has come to the fore for the Labour Party now because the New Labour project was contradictory, 'containing elements that were simultaneously metropolitan liberal and communitarian; patriotic and internationalist; Fabian and devolutionist; followers of economic liberalism in the financial sector and tax and spend social democrats' (Beech and Hickson 2014, 84).

Blair's refashioning of the Party was electorally successful but in the 'process of "modernisation" the Labour party gradually lost the intellectual distinctiveness of its creed' (Beech 2009, 529). The linchpin of the New Labour project was that macro-economic growth would be the remedy for fostering equality and facilitating meritocratic liberty (Finlayson 2013). Yet the Party ultimately failed to recognise that growth is not distributed equally and that a sense of collective community, solidarity and a shared albeit constructed national identity (Anderson 2006), which immigration is seen by some to threaten, matter to a great deal of the electorate.

Labour's immigration policy is a fantastic example of this—the pursuit of macro liberal policies for what they believed was good for the economy and therefore the country, but which generated rapid change in communities. Yet Labour did not address or consider the impact on the local level of these changes nor were its effects on social cohesion and communities necessarily thought through. This incidentally also exemplifies a wider institutional problem, where the tax benefits of immigration are slow to trickle down and compensate organisations such as the NHS or local education authorities who feel the strain for the costs of providing additional public services required by any increase in population (Tilford 2015, 3; House of Lords 2008).

Labour face a dilemma on immigration, but this goes beyond the policy direction and strategy of immigration specifically. The dilemma it faces is an exemplification of the core dilemma for the centre left in today's political context—it is a wider identity crisis, and one without easy answers, about who the Party represents and what its ideological purpose is. Former Labour Home Secretary Roy Hattersely puts his finger on the deep-seated nature of the problem commenting on the 2010 defeat: 'The Party not only failed to set out a clear and coherent idea of what it proposed to do. It was not even sure about the purpose of its existence' (Roy Hattersley 2010 quoted in Hain 2016, p. V). This diagnosis resonates even more seven years later.

As we saw in Chaps. 2 and 5, there is no fixed path ideologically for the centre left on immigration, but this tells us something wider about the identity of the Labour Party and the centre left more generally. It points to a lack of, or at the very least confidence in, a core, coherent ideology and hints at the underlying uneasiness between internationalism and nationalism (Beech 2016). On the one hand, one can couch open borders and free movement as emblematic of neoliberalism, globalisation and the other capitalist forces that exploit those on low incomes. On this basis, the Labour Party could comfortably advocate restrictions on immigration in line with their protectionist core, as they have done previously in the postwar era, and to some extent under Miliband. On the other hand, challenging the populist right rhetoric, challenging racism and discrimination, and defending the prized notion of class solidarity blind to nationality are equally the tasks for Labour if they are to stand for anything. The conundrum for Labour is that they are torn between a worldview that emphasises internationalism 'and the magnetic pull of electorates towards enduring ties of nationhood, identity and belonging' (Diamond 2014). Labour is not alone in this battle; centre-left parties across the Western world are struggling to fashion concrete answers, which is ultimately symptomatic of how or even whether social democracy can be revived in today's political context (Hickson 2016; Ryner 2010).

So immigration cuts to the core of the Party and their uneasiness or at least indecision between these two very different representational bases, which require very different policies, and fundamentally the challenge to carve a narrative that chimes with the core foundations of what Labour once represented. The Labour government's managed migration policy has highlighted a deep division in Labour's electoral bases and in turn brought to the fore the ideological tension of the Labour Party. These tensions were always extant, yet with immigration only recently being top of the political agenda, are only now visibly polarising debate. This is one of the key legacies of the New Labour period and may have a detrimental effect on the Party's future as a governing party.

8.3.2 New Labour's Legacies: The Backlash Against Globalist Policies

To get to the essence of Labour's legacy on the politics of immigration, we conclude by returning to one of the core arguments of the book, the central explanation for this case of policy change and the subject of Chap. 5—New Labour's globalist ideology. New Labour's reforms were deep seated in the belief that globalisation was both inevitable and intrinsically positive. The referendum result marked a seismic shift in British politics, bringing into stark relief a new fault line that has been developing for a number of years,

> Pitting "liberal cosmopolitans" against those with more traditional values [reflecting] the divide between the winners and losers from the social and economic changes that have taken place over the past several decades – deindustrialisation, globalisation and digitisation. (Birch 2016, 107)

The leave side was especially strong in the Labour heartlands of post-industrial north-eastern towns with larger working class populations (Hobolt 2016, 1273) where deindustrialisation struck hardest and where average incomes have stagnated (Coyle 2016, 23). These are areas that have felt the pangs of rapid change brought about by globalist policies, such as global offshoring, outsourcing to developing nations with minimal labour standards, the complete decline of manufacturing sectors, and the more general post-industrial shift to high technology and a broader service sector. 'Post-traumatic cities', as Justin Gest calls them, where urban communities have lost signature industries in the mid to late twentieth century and never really recovered (Gest 2016, 7), in turn bringing unemployment, alienation and a loss of agency for many. In this sense, the referendum campaign 'ended up becoming, to a very large extent, a debate about globalisation in its local, European manifestation' (O'Rouke 2016, 45). As we have seen, immigration is the human face of globalisation; hence, the issue is front and centre of the political landscape. One could interpret the referendum result then as the rejection of New Labour's branded cosmopolitan, cool Britannia, wrapped up in a happy vision of globalisation (Calhourn 2016).

For New Labour, with Blair at the helm, marshalled the narrative that there simply was no alternative to the relentless forces of globalisation and that people must adapt and move with the times or risk decline. Take Blair's 2005 leadership speech, pitting those who embraced change against those who fear it:

> The pace of change can either overwhelm us, or make our lives better and our country stronger. What we can't do is pretend it is not happening. I hear people say we have to stop and debate globalisation. You might as well debate whether autumn should follow summer...The character of this changing world is indifferent to tradition. Unforgiving of frailty. No respecter of past reputations. It has no custom and practice...It is replete with opportunities, but they only go to those swift to adapt, slow to complain, open, willing and able to change. (Blair 2005)

Blair went so far as to suggest that there was no other kind of politics or political economy suitable for the times: 'In the era of rapid globalisation, there is no mystery about what works: an open, liberal economy, prepared constantly to change to remain competitive' (Blair 2005). Posing globalisation and what Blair deemed the appropriate policy responses to embrace the so-called opportunities of it as inevitable was clearly a winning electoral strategy in the 2000s. Yet this polarisation seemed to ignore the publics' desire for control, security and social order. Thus when the financial crisis hit in 2008—without doubt at least propelled by the highly globally interdependent financial systems New Labour had helped create—it is little wonder why much of the British public felt let down, alienated and angry at this apparently inevitably global, open system for which many were certainly not seeing the benefits, only the costs.

We should be wary of presentism and single causation though; few scholars would claim a single cause for the unquestionable discontent with the status quo generally, and immigration is just one, albeit important, facet of general distrust with the elite and anti-establishment feeling (McLaren 2012). Nonetheless, the way Blair spoke about the global, liberal world as unstoppable gave publics little choice for any alterative visions, and thus, when the system failed them, a backlash against this type of politics was foreseeable. One would be hard pressed to claim that New Labour was the cause of this rupture, but their policies brought this fault line to the fore. That the New Labour government pursued an expansive immigration regime to dovetail with their globalised political economy only exacerbated this. Immigration is simply the most concrete representation of globalisation and therefore the easiest target for public antagonism in contrast to the faceless and abstract forces of neoliberalism, globalisation and austerity.

New Labour's managed migration policy has generated many legacies. The reforms brought hundreds of thousands of immigrants, which have economically benefited the country, culturally enriched society for generations to come and 'literally changed the face of Britain' (Finch and Goodhart 2010). Nevertheless, the regime was undoubtedly electorally costly for the Labour Party, and their legacy plays a part in the story of the politicisation of immigration. Perhaps the greatest legacy is that no government would dare to pursue an expansive immigration regime again, and in this sense, policy has paved a new way for politics. At the very least, New Labour's policies brought immigration to the fore of the political landscape and perhaps solidified it as a permanent fixture. In this sense, it is no exaggeration to say that the period under Labour has transformed the politics of immigration in Britain.

REFERENCES

Anderson, B. (2006). *Imagined Communities: Reflections on the Origin and Spread of Nationalism*. London: Verso Books.

Ashcroft, L. (2015). *Ashcroft National Poll: Con 34%, Lab 31%, Lib Dem 7%, UKIP 14%, Green 7%*. Available from: http://lordashcroftpolls.com/2015/03/ashcroft-national-poll-con-34-lab-31-lib-dem-7-ukip-14-green-7/

Bale, T. (2008). Turning Round the Telescope. Centre-Right Parties and Immigration and Integration Policy in Europe. *Journal of European Public Policy, 15*(3), 315–330.

Bale, T. (2014). Putting It Right? The Labour Party's Big Shift on Immigration Since 2010. *The Political Quarterly, 85*(3), 296–303.

Bale, T. (2015). *Five Year Mission: The Labour Party Under Ed Miliband*. Oxford: Oxford University Press.

Bale, T., Green-Pedersen, C., Krouwel, A., Luther, K. R., & Sitter, N. (2010). If You Can't Beat Them, Join Them? Explaining Social Democratic Responses to the Challenge from the Populist Radical Right in Western Europe. *Political studies, 58*(3), 410–426.

BBC. (2006, September 6). *Cameron Criticises Globalization* [Online]. Available from: http://news.bbc.co.uk/1/hi/uk_politics/5318636.stm. Accessed 20 Dec 2016.

BBC. (2017, January 7). *Theresa May: UK Cannot Keep 'Bits' of EU Membership* [Online]. http://www.bbc.co.uk/news/uk-politics-38546820. Accessed 8 Jan.

Beech, M. (2009). No New Vision: The Gradual Death of British Social Democracy? *The Political Quarterly, 80*(4), 526–532.

Beech, M. (2016). Internationalism. In K. Hickson (Ed.), *Rebuilsing Social Democracy* (pp. 127–141). Bristol: Policy Press shorts insights.

Beech, M., & Hickson, K. (2014). Blue or Purple? Reflections on the Future of the Labour Party. *Political Studies Review, 12*(1), 75–87.

Birch, J. (2016). Our New Voters Brexit, Political Mobilisation and the Emerging Electoral Cleavage. *IPPR Juncture, 23*(2), 107–110.

Blair, T. (1994). *Leader's Speech, Blackpool* [Online]. Available from: http://www.britishpoliticalspeech.org/speech-archive.htm?speech=200. Accessed 20 Dec 2016.

Blair, T. (2005). *Keynote Speech to Labour Party Conference, Brighton* [Online]. Available at http://news.bbc.co.uk/1/hi/uk_politics/4287370.stm. Accessed 20 Dec 2016.

Bogdanor, V. (2016). A New Way Back for Social Democracy?: The EU Referendum and Its Lessons for the Left. *IPPR Juncture, 23*(1), 8–11.

Cabinet Office. (2011). *Public Bodies Reform – Proposals for Change*. London: Cabinet Office. Available at: https://www.gov.uk/government/uploads/system/uploads/attachment_data/file/62125/Public_Bodies_Reform_proposals_for_change.pdf. Accessed 23 Aug 2013.

Calhourn, C. (2016). Brexit Is a Mutiny Against the Cosmopolitan Elite. *New Perspectives Quarterly, 33*(3), 50–58.

Cameron, D. (2014, November 28). *JCB Staffordshire: Prime Minister's Speech* [Online] https://www.gov.uk/government/speeches/jcb-staffordshire-prime-ministers-speech. Accessed 18 Dec 2016.

Cameron, D. (2015, May 21). *PM Speech on Immigration* [Online]. https://www.gov.uk/government/speeches/pm-speech-on-immigration. Accessed 20 Dec 2016.

Caviedes, A. (2009). *Prying Open Fortress Europe: The Turn to Sectoral Labor Migration*. Lanham: Lexington Books.

CIPD [Chartered Institute of Personnel and Development]. (2013). *The State of Migration: Employing Migrant Workers*. London: CIPD.
Clarke, S., & Whittaker, M. (2016). *The Importance of Place: Explaining the Characteristics Underpinning the Brexit Vote Across Different Parts of the UK*. London: Resolution Foundation.
Conservative Party. (2010). *Conservative Party General Election Manifesto: Invitation to Join the Government of Britain*. London: Conservative Party.
Conservative Party. (2017). *Forward Together: The Conservative Manifesto* [Online]. https://www.conservatives.com/manifesto. Accessed 3 June 2017.
Cooper, Y. (2014, April 10). Immigration Speech. *Labourlist* [Online]. http://labourlist.org/2014/04/yvette-coopers-immigration-speech-full-text/. Accessed 24 Oct 2014.
Corbyn, J. (2016, September 28). Speech to the Party of European Socialists Council in Prague. *Labourlist* [Online] http://press.labour.org.uk/post/153983095494/jeremy-corbyn-speech-to-the-party-of-european. Accessed 13 Jan 2017.
Corbyn, J. (2017, January 10). *Speech in Peterborough*. Labourlist http://press.labour.org.uk/post/155675139454/jeremy-corbyns-speech-in-peterborough. Accessed 12 Jan 2017.
Cowburn, A., & The Independent. (2017, January 4). *Andrea Leadsom Hints Farmers Could Continue to Hire EU Migrants After Brexit for Seasonal Work* [Online]. http://www.independent.co.uk/news/uk/politics/andrea-leadsom-hints-farmers-could-continue-to-hire-eu-migrants-after-brexit-for-seasonal-work-a7509906.html. Accessed 6 Jan.
Cowley, P., & Kavanagh, D. (2016). *The British General Election of 2015*. London: Palgrave Macmillan.
Coyle, D. (2016). Brexit and Globalisation. In R. E. Baldwin (Ed.), *Brexit Beckons: Thinking Ahead by Leading Economists* (pp. 23–39). London: CEPR press.
Cruddas, J. (2016). *What We Can Learn from Labour's Crushing Election Defeat*. Labourlist [Online]. http://labourlist.org/2016/05/labours-future-what-we-can-learn-from-the-election-loss/. Accessed 13 Jan 2017.
Curtice, J. (2016). Not the End, My Friend Labour, the Referendum Result and the Challenges that Remain. *IPPR Juncture, 23*(1), 18–21.
Dahlgreen, W. (2013). *Voters: Immigration Was Too High Under Labour*. London: YouGov [Online] https://yougov.co.uk/news/2013/10/15/voters-immigration-too-high-labour/
Dennison, J., & Goodwin, M. (2015). Immigration. Issue Ownership and the Rise of UKIP. In A. Geddes & J. Tonge (Eds.), *Britain Votes 2015* (pp. 168–188). Oxford: Oxford University Press.
Diamond, P. (2014 June 4). Social Democracy Is on the Ropes – It Needs a New Vision. *The Guardian* [Online] https://www.theguardian.com/commentisfree/2014/jun/04/social-democracy-on-ropes-new-vision-labour. Accessed 20 Dec 2016.

Duffy, B. (2014). Perceptions and Reality: Ten Things We Should Know About Attitudes to Immigration in the UK. *The Political Quarterly, 85*(3), 259–266.

Evans, S. (2008). Consigning Its Past to History? David Cameron and the Conservative Party. *Parliamentary Affairs, 61*(2), 291–314.

Evans, G., & Chzhen, K. (2013). Explaining Voters' Defection from Labour over the 2005–10 Electoral Cycle: Leadership, Economics and the Rising Importance of Immigration. *Political Studies, 61*(1 suppl), 138–157.

Fabian Society. (2017). *Stuck: How Labour Is Too Weak to Win, and Too Strong to Die.* London: Fabian Society [Online]. Available from: http://www.fabians.org.uk/wp-content/uploads/2016/12/Stuck-Fabian-Society-analysis-paper.pdf. Accessed 31 Jan 2017.

Finch, T., & Goodhart, D. (2010). Introduction. In T. Finch & D. Goodhart (Eds.), *Immigration Under Labour* (pp. 3–10). London: IPPR.

Finlayson, A. (2013). From Blue to Green and Everything in Between: Ideational Change and Left Political Economy After New Labour. *British Journal of Politics and International Relations, 15*(1), 70–88.

Ford, R. (2011). Acceptable and Unacceptable Immigrants: How Opposition to Immigration in Britain Is Affected by Migrants' Region of Origin. *Journal of Ethnic and Migration Studies, 37*(7), 1017–1037.

Ford, R., & Goodwin, M. J. (2014). *Revolt on the Right: Explaining Support for the Radical Right in Britain.* London: Routledge.

Ford, R., Jennings, W., & Somerville, W. (2015). Public Opinion, Responsiveness and Constraint: Britain's Three Immigration Policy Regimes. *Journal of Ethnic and Migration Studies, 41*(9), 1391–1411.

Gamble, A. (2014). Austerity as Statecraft. *Parliamentary Affairs, 68*(1), 42–57.

Geddes, A. (2014). The EU, UKIP and the Politics of Immigration in Britain. *The Political Quarterly, 85*(3), 289–295.

Geddes, A., & Tonge, J. (2015). *Britain Votes 2015.* Oxford: Oxford University Press.

Gest, J. (2016). *The New Minority: White Working Class Politics and Marginality.* Oxford: Oxford University Press.

Gower, M. (2015). *Immigration and Asylum: Changes Made by the Coaltiion Government 2010–2015.* London: House of Commons Library.

Green, D. (2012). *Damian Green's Speech on Making Immigration Work for Britain.* Delivered at the Policy Exchange on 2 February 2012.

Green, J., & Hobolt, S. B. (2008). Owning the Issue Agenda: Party Strategies and Vote Choices in British Elections. *Electoral Studies, 27*(3), 460–476.

Hain, P. (2016). Foreword: Rediscovering Confidence and Soul. In K. Hickson (Ed.), *Rebuilding Social Democracy* (pp. v–1). Bristol: Policy Press Shorts Insights.

Hardman, I. (2014, January 11). *Ed Miliband's Immigration Nightmare, Spectator.* https://www.spectator.co.uk/2014/01/labours-immigration-nightmares/

Hickson, K. (2016). *Rebuilding Social Democracy: Core Principles for the Centre Left*. Bristol: Policy Press.
Hobolt, S. (2016). The Brexit Vote: A Divided Nation, a Divided Continent. *Journal of European Public Policy, 23*(9), 1259–1277.
Home Office. (2015). *Policy Paper: 2010 to 2015 Government Policy: Immigration and Borders*. London: HM Government.
House of Lords. (2008). *The Economic Impact of Immigration. House of Lords Select Committee on Economic Affairs, 1st Report of session 2007–8*. London: Stationery Office.
Inglehart, R. F. (2008). Changing Values Among Western Publics from 1970 to 2006. *West European Politics, 31*(1–2), 130–146.
Ipsos-Mori. (2015). *Issues Facing Britain: August 2015 Issues Index*. Available from: https://www.ipsos-mori.com/researchpublications/researcharchive/3614/EconomistIpsos-MORI-August-2015-Issues-Index.aspx. Accessed 15 Sep 2015.
Ipsos-Mori. (2017). *MPs Winter Survey 2016 Key Influencer Tracking*. Report Prepared for Kings College London and Queen Mary University of London.
Ivarsflaten, E. (2005). Threatened by Diversity: Why Restrictive Asylum and Immigration Policies Appeal to Western Europeans. *Journal of Elections, Public Opinion & Parties, 15*(1), 21–45.
Kaufman, E. (2016). *It's NOT the Economy, Stupid: Brexit as a Story of Personal Values'*. LSE: British Politics and Policy Blog [Online]. Accessed 20 Jan.
Kaufman, E., & Harris, G. (2014). *Changing Places: Mapping the White British Response to Ethnic Change*. London: Demos.
Leach, R. (2015). *Political Ideology in Britain*. Basingstoke: Palgrave Macmillan.
McLaren, L. (2012). Immigration and Trust in Politics in Britain. *British Journal of Political Science, 42*(1), 165–185.
Miliband, E. (2012, June 22). *New Immigration Requires a New Economy*. Speech Given to the IPPR at the Royal Festive Hall. Available from: http://www.politics.co.uk/comment- analysis/2012/06/22/ed-miliband-s- immigration-speech-in-full. Accessed 14 Mar 2013.
Miliband, E. (2012a, December 14). *Ed Miliband Immigration Speech*. Tooting. Available from: http://www.newstatesman.com/staggers/2012/12/full-text-ed-miliband-immigration-speech. Accessed 14 Mar 2013.
Miliband, E. (2012b, June 22). *New Immigration Requires a New Economy*. Speech Given to the IPPR at the Royal Festive Hall. Available from: http://www.politics.co.uk/comment-analysis/2012/06/22/ed-miliband-s-immigration-speech-in-full. Accessed 14 Mar 2013.
Moynihan, D. P., & Soss, J. (2014). Policy Feedback and the Politics of Administration. *Public Administration Review, 74*(3), 320–332.
Norris, P. (2005). *Radical Right: Voters and Parties in the Electoral Market*. Cambridge/New York: Cambridge University Press.

O'Rouke, K. H. (2016). This Backlash Has Been a Long Time Coming. In R. E. Baldwin (Ed.), *Brexit Beckons: Thinking Ahead by Leading Economists* (pp. 43–52). London: CEPR press.

Odmalm, P. (2014). *The Party Politics of the EU and Immigration.* London: Palgrave Macmillan.

Portes, J. (2016). Immigration – The Way Forward. In R. E. Baldwin (Ed.), *Brexit Beckons: Thinking Ahead by Leading Economists* (pp. 105–111). London: CEPR press.

Ryner, M. (2010). An Obituary for the Third Way: The Financial Crisis and Social Democracy in Europe. *The Political Quarterly, 81*(4), 554–563.

Schattschneider, E. E. (1935). *Politics, Pressures and the Tariff.* New York: Prentice-Hall.

Skocpol, T. (1992). *Protecting Soldiers and Mothers: The Political Origins of Social Policy in the United States.* Cambridge: Belknap Press of Harvard University Press.

Telegraph. (2013, November 13). Labour Made a "Spectacular Mistake" on Immigration, Admits Jack Straw. *Telegraph.* Available from: http://www.telegraph.co.uk/news/uknews/immigration/10445585/Labour-made-aspectacular-mistake-on-immigration-admits-Jack-Straw.html. Accessed 7 Jan 2014.

Tilford, S. (2015). Britain, Immigration and Brexit. In *Bulletin Article.* London: CEPR press.

Appendix 1: Overview of Economic Immigration Policy Reforms and Rhetoric, 1997–2010

Year	Economic immigration	Asylum, integration, family, border controls and irregular	(a) Home secretary (b) Minister for immigration
1997		Primary purpose rule dropped Opt out of Amsterdam Treaty	a. Jack Straw b. Michael O'Brien
1998	DTI White Paper Building the Knowledge Driven Economy	HO White Paper *Fairer, Faster, Firmer* Human Rights Act	a. Jack Straw b. Michael O'Brien
1999	Prime Minister's Initiative on International Education	Immigration and Asylum Act	a. Jack Straw b. Michael O'Brien/ Barbara Roche
2000	Innovators Scheme introduced Barbara Roche speech Work permit criteria reduced	Race Relations Amendment Act	a. Jack Straw b. Barbara Roche
2001	Work Permits UK moves to the Home Office SAWS quota increased to 15,200 places HO and PIU research paper *Migration: an economic and social analysis* published		a. Jack Straw/ David Blunkett b. Barbara Roche/ Lord Rooker

(*continued*)

(Continued)

Year	Economic immigration	Asylum, integration, family, border controls and irregular	(a) Home secretary (b) Minister for immigration
2002	HO White Paper *Secure border, safe havens* published HSMP introduced	Sangatte crisis Nationality, Immigration and Asylum Act	a. David Blunkett b. Lord Rooker/ Beverley Hughes
2003	SAWS quota increased to 25,000 SBS created Working Holiday Makers Scheme liberalized		a. David Blunkett b. Beverely Hughes
2004	A8		a. David Blunkett/ Charles Clarke b. Beverely Hughes/ Des Browne
2005	HO five year departmental plan *Controlling our borders: making migration work for Britain* published		a. Charles Clarke b. Des Browne/Tony McNulty
2006	Prime Minister's Initiative on International Education renewed for another six years (PMI2) HO policy strategy *A Points-based system: making migration work for Britain* published SBS closed	Reid declares IND 'unfit for purpose'	a. Charles Clarke/John Reid b. Tony McNulty/Liam Byrne
2007	A2 MAC and MIF created	UK Borders Bill	a. John Reid/Jacqui Smith b. Liam Byrne
2008	Points-based system fully operational	IND merges with HM Revenues and Customs to become UK Border Agency	a. Jacqui Smith b. Phil Woolas
2009	Tier 1 criteria tightened		a. Jacqui Smith/Alan Johnson b. Phil Woolas
2010	Brown states 'British jobs for British workers'		a. Alan Johnson b. Phil Woolas

Index

A
A2 accession, 72, 75
A8 accession, 74, 75, 110
asylum, 4, 5, 7, 28, 62–7, 76, 122, 134, 135, 137, 138, 140, 141, 143–51, 165, 171, 180, 194, 206, 208, 209, 220
Australia, 190

B
Balch, Alex, 2, 4, 28, 66–8, 76, 78, 87, 102, 103, 135, 174, 179, 191, 192, 203, 204
Bale, Tim, 3, 8, 10, 30, 32–6, 64, 119, 122, 123, 137, 142, 144, 145, 148–51, 208, 219, 227, 228, 230
Blair, Tony, 64, 73, 78, 106, 107, 110, 122–7, 129–32, 134, 137, 140, 141, 143, 148–51, 165, 166, 168–70, 173, 176, 178, 180, 187, 192, 212, 220, 224, 232, 235, 236
Blunkett, David, 67, 103, 105, 111, 129, 132, 133, 135, 143, 145, 162, 164, 165, 173, 178–82, 185–7, 192, 194, 201, 203, 208, 210
Boswell, Christina, 11, 21, 107, 113, 146, 167, 188–93, 210
Brexit, 1, 222, 224
British Chamber of Commerce (BCC), 91, 93, 97, 108, 109, 148
British Nationality Act of 1948, 181
British National Party (BNP), 122, 139, 148, 151, 226
Brown, Gordon, 77, 78, 93, 97, 98, 106, 107, 124, 127, 129, 130, 137, 140, 141, 143, 169, 170, 176, 194, 209, 212, 219
Byrne, Liam, 79, 112, 173, 190

C
Cabinet Office, 166, 167, 171, 223
Callaghan, James, 57, 164
Cameron, David, 146, 220, 224, 226–8
Canada, 190
Carriers Liability Act of 1987, 63, 134
Caviedes, Alexander, 8, 9, 27, 28, 64, 71, 87, 91, 98, 99, 101, 223

© The Author(s) 2018
E. Consterdine, *Labour's Immigration Policy*,
https://doi.org/10.1007/978-3-319-64692-3

census, 6, 140, 201
Central and Eastern Europe (CEEC), 147, 178, 224. *See also* A8 accession
centre left parties, 34–6, 122, 123, 152, 204, 205, 229, 230, 233. *See also* Labour Party
centre right parties, 33, 34, 142, 152. *See also* Conservative Party
Clarke, Charles, 63, 71, 72, 100, 108, 177, 220
coalition government (2010–2015), 110, 142, 219, 223
commonwealth, 54–6, 58–62, 65, 72, 99
Commonwealth Immigration Act of 1962, 56, 121
Commonwealth Immigration Act of 1968, 121
Confederation for British Industry (CBI), 90, 91, 93, 97, 99, 100, 102, 104–8, 110–12, 175
Conservative Party, 34, 58, 60, 62, 77, 123, 142, 144–6, 148, 149, 152, 204, 206, 219, 221, 222
constructivism, 38, 39, 93, 209. *See also* historical institutionalism; institutions
Corbyn, Jeremy, 229–31
cosmopolitanism, 128
critical junctures, 37, 212
Cruddas, John, 3, 126, 128, 139, 140, 229

D
Demos, 97
Denham, John, 132, 139–41, 177, 181
Department for Business and Innovation (BIS), 170
Department for Communities and Local Government (DCLG), 182

Department for Education and Employment (DfEE), 70, 101, 103, 175–7, 185, 186, 191, 194
Department for Trade and Industry (DTI), 68, 175
Department for Work and Pensions (DWP), 74, 169
departmentalism, 162–5

E
European Commission, 66
European Union (EU), 1, 2, 5, 34, 43, 44, 66, 72, 74, 75, 78, 91, 93, 124, 125, 138, 147, 148, 178, 181, 182, 221, 222, 224–31
evidence-based policymaking, 162, 166–8, 189, 190, 193, 194

F
Fairer, Faster, Firmer (1998 White Paper), 66, 144
Farage, Nigel, 225, 226, 230
feedback effects, 14, 42, 43, 221–4
financial crisis or crash, 76, 219, 224, 235
Ford, Rob, 34, 139, 225, 226
Foreign and Commonwealth Office (FCO), 45, 177–81
Freeman, Gary, 2, 3, 6, 8–10, 13, 25, 26, 53, 64, 87, 98, 113, 204, 227
Fresh Talent scheme working in Scotland, 100

G
Geddes, Andrew, 3, 27, 28, 58, 66, 77, 96, 121, 136, 149, 204, 219, 224

General Election
  1997, 122, 142, 144, 211
  2001, 144, 145, 150
  2005, 76, 77, 150, 206, 207
  2010, 3
  2015, 219
  2017, 221
Giddens, Anthony, 124, 133
globalisation, 4, 33, 128–31, 134, 138, 153, 202, 205–7, 212, 227, 233–6
Goodhart, David, 2, 65, 144, 236
Goodwin, Matthew, 1, 34, 139, 220, 225, 226
Gould, Philip, 77, 78, 150

H
Hall, Peter, 22, 23, 40, 53, 80, 108, 185, 187, 190, 193, 204, 209
Hampshire, James, 5, 8, 9, 42, 43, 53, 54, 56, 59, 99, 119, 121, 134, 208, 213
Hansen, Randall, 2, 3, 11, 43, 53, 54, 57, 58, 62, 64, 121, 181, 213
highly skilled migrants programme (HSMP), 68, 69, 71, 78, 100, 170, 171
historical institutionalism, 9, 11, 13, 37–44, 161, 202
Hodge, Margaret, 130, 132, 139, 140, 175–7
Hollifield, James, 2, 5, 6, 9, 75
Home Office, 5, 45, 57, 60, 66, 69–76, 79, 95, 99, 101, 103, 105–7, 109, 111, 112, 114, 129, 135, 141, 150, 162, 168–84, 186–92, 194, 195, 208, 209, 211–13, 222
House of Commons, 55, 143, 144, 151
House of Lords, 55, 173, 233
Hughes, Beverley, 75, 134

I
ideology, 10, 13, 29–37, 119–53, 162, 194, 195, 202, 204, 207, 212–14, 232–4
Immigration Act of 1971, 59, 61, 121
Immigration Nationality Directorate (IND), 71, 75, 76, 79, 108, 183
Innovators Scheme, 68, 100, 170
Institute for Public Policy Research (IPPR), 69, 70, 94, 95, 97, 102–5, 132, 143, 191, 203, 210
Institute of Directors (IoD), 91, 97
institutions, 11, 12, 21, 22, 37–41, 43, 44, 66, 73, 111, 132, 161, 162, 176, 194, 195, 201, 208–10, 221. See also historical institutionalism
interest groups, 9, 13, 21, 24–7, 29, 30, 42, 45, 87–93, 96, 97, 99, 100, 105, 106, 109, 111–14, 162, 202–4, 213, 221. See also organised interests
international students, 73, 74, 76, 176, 177, 186. See also points-based system, Tier 4 students; Prime Ministers Initiative on International Education (PMI1 and PMI2)
intra-company transfers (ICT), 70, 71, 98, 101, 223
intra-party divisions, 34
Ipsos Mori, 6, 136, 152, 220, 221. See also public opinion on immigration

J
joined-up government (JUG), 13, 166, 168–87, 194

K

Kingdon, John, 6, 28, 40, 41, 43, 103, 203, 210, 211
Kinnock, Neil, 122–4

L

Labour Party, 30, 37, 59, 65, 77, 78, 93–5, 97, 99, 102, 106, 107, 114, 120–2, 126–9, 131–3, 136–8, 140, 143, 150, 153, 162, 170, 179, 202, 203, 205–8, 210, 211, 225, 228, 230, 232–4, 236. *See also* New Labour
Letwin, Oliver, 142–52
Liberal Democrats, 146, 226

M

Mandelson, Peter, 125, 175
Mattison, Deborah, 77, 150
May, Theresa, 1, 221
Menz, Georg, 8, 9, 27, 28, 87, 90, 91, 96, 99, 104, 107, 108, 114, 130
methods/methodology, 12, 22, 44–6, 148, 173
Migrant Impacts Forum (MIF), 112, 113, 182
Migrant Rights Network (MRN), 104
Migration Advisory Committee (MAC), 74–6, 78, 79, 112, 113, 172, 189, 223, 224
Migration Watch (MW), 94–7
multiculturalism, 44, 132, 205
Mulvey, Gareth, 68, 77, 135, 143, 209

N

National Farmers Union (NFU), 91, 97, 101, 102, 105, 110. *See also* trade unions; organised interests; interest groups
National Health Service (NHS), 78, 182, 233
neoliberalism, 207, 233, 236
New Labour, 2, 3, 14, 36, 46, 64, 65, 94, 95, 97, 106, 107, 120–6, 128, 129, 131, 134, 143, 153, 165–8, 172, 203, 212, 219–36
New Public Management (NPM), 94, 189
Nuttall, Paul, 226

O

Office for National Statistics (ONS), 228
Organisation for Economic Co-operation and Development (OECD), 74, 75, 180, 182, 205
organised interests, 9, 10, 12, 13, 21, 23–9, 87–114, 130, 161, 188, 201–4, 213, 221. *See also* interest groups

P

Parliamentary Labour Party (PLP), 136–8, 141, 152, 206
path dependency, 38, 42
Pearce, Nick, 103, 104, 135, 136, 181, 203
Performance and Innovation Unit (PIU), 70, 168, 173, 190–2
points-based system
    Tier 1 general, 222
    Tier 1 post-study, 222
    Tier 2 genera, 222
    Tier 4 students, 76
    Tier 5 temporary workers, 76

policy transfer, 187–93
Powell, Enoch, 58
Prime Ministers Initiative on International Education (PMI1 and PMI2), 176
prisoners scandal, 78
public opinion on immigration, 10, 26, 213, 220. *See also* Ipsos Mori
Public Service Agreements (PSA), 170, 171, 175

R
race relations, 44, 53, 57, 61, 64, 66, 121, 122, 144, 208, 210
rebellions, 32, 136–8
referendum–2016 EU referendum, 1
Refugee Council, 104, 164
Reid, John, 78
Roche, Barbara, 68, 69, 100, 128, 171

S
Saggar, Shamit, 7, 53, 58, 61, 77, 131, 136, 144, 152
Schmidt, Vivien, 38
Seasonal Agricultural Workers Scheme (SAWS), 71, 101, 222, 224
Sector Based Scheme (SBS), 71, 78, 100, 101, 222
Somerville, Will, 27, 65, 68, 69, 87, 91, 99, 112, 141, 150
Soysal, Yasmin, 4
special advisors (SpAds), 45, 100, 101, 103, 135, 186, 203
Spencer, Sarah, 59, 66, 71, 76, 78, 93, 102, 103, 175, 176, 191, 203
spillover effects, 166
Statham, Paul, 27, 28, 31, 96, 204
Strategy Unit, 168, 172, 174, 183, 190–2
Straw, Jack, 138, 144, 178–80, 185, 186, 225

T
Thatcher, Margaret, 62, 124, 133, 134, 143, 148, 201, 226
third way, 120, 124, 126–31, 162, 166, 194, 195, 202, 205, 212, 213, 232
Trade Union Congress (TUC), 90, 92, 93, 97, 101, 106, 107, 109, 110, 112, 125. *See also* trade unions
trade unions, 45, 92, 93, 105, 107, 124. *See also* National Farmers Union (NFU); Trade Union Congress (TUC); Unison; Unite
Treasury, 45, 59, 60, 68, 100, 101, 138, 162, 169–75, 177, 181–4, 186, 191, 192, 194, 209–12

U
UK Border Agency (UKBA), 79, 108, 164, 165
UK Independence Party (UKIP), 151, 225–31
UK Visas and Immigration (UKVI), 45, 79, 222, 223
Unison, 92, 97, 107. *See also* trade unions
Unite, 137. *See also* trade unions

W
Whitehall, 54, 67, 134, 153, 162–4, 167, 168, 170, 178, 191, 192
window of opportunity, 40–3, 103, 210–14. *See also* Kingdon, John
Work Permits UK, 70, 71, 74, 101, 146
working holidaymakers/youth mobility scheme, 71, 72, 76, 79

Printed by Printforce, the Netherlands